SpringerBriefs in Earth Sciences

More information about this series at http://www.springer.com/series/8897

Elena Tolkova

Tsunami Propagation in Tidal Rivers

 Springer

Elena Tolkova
NorthWest Research Associates, Inc
Kirkland, WA, USA

ISSN 2191-5369   ISSN 2191-5377   (electronic)
SpringerBriefs in Earth Sciences
ISBN 978-3-319-73286-2   ISBN 978-3-319-73287-9   (eBook)
https://doi.org/10.1007/978-3-319-73287-9

Library of Congress Control Number: 2017962901

Printed on acid-free paper

This Springer imprint is published by Springer Nature
The registered company is Springer International Publishing AG
The registered company address is: Gewerbestrasse 11, 6330 Cham, Switzerland

Foreword

Flooding from the sea took thousands of lives in northeast Japan in December 1611. News of the disaster reached the first Tokugawa shogun. One of his aides wrote of the flooding as a "so-called tsunami"—an early conjoining of characters for harbor (*tsu*) and wave or waves (*nami*).

It is unlikely that the 1611 tsunami was confined to harbors. In northeast Japan it reached heights of 5 m or more. So did the disastrous 2011 Tohoku tsunami—and its waves, in addition to flooding coastal plains, ran into and out of rivers.

Coastal rivers have also been invaded by tsunamis from the eastern Pacific. The main Chilean tsunami of 1960 swept away riverside villages near its source. Near-field tsunamis in North America's Cascadia region are evidenced by sheets of sand beside tidal streams that successive waves ascended. Documented far-field effects of the 1700 Cascadia tsunami include anomalous river-mouth waves, in northeast Japan, that occasioned the loss of a rice freighter and two of its crew.

In this book, twenty-first century measurements provide starting points for exploring how tsunamis interact with rivers. The measurements were made in rivers on both sides of the Pacific and beside the Indian Ocean as well. They inform analyses of how a riverine tsunami increases average water levels, why it intensifies during rising tides, and how it behaves after transforming into bores. The findings shed light on the fluvial expressions of the so-called harbor waves.

Seattle, WA, USA   Brian F. Atwater
Scientist Emeritus, U.S. Geological Survey
Affiliate Professor, University of Washington
October 2017

Preface

"One cannot step into the same river twice, for other waters are continually flowing on" (Heraclitus of Ephesus, 535 BC – 475 BC).

People have settled around rivers for probably as long as they have inhabited the planet Earth. To most, the rivers are one-way systems, where the waters make their way toward the ocean. The river water is like time: it is always moving in one direction, from past to the future, from mountains to the ocean. Closer to the ocean, though, the river's philosophy changes. The ocean continuously invades rivers with tides, and occasionally—with storm surges and tsunamis. The amount of water entering the river with rising tide and leaving it with falling (ebb) tide—a so-called tidal prism—can exceed the freshwater inflow from the upstream by several times. The amount of water brought into a river by a large tsunami can exceed the tidal prism by several times. Next to the mouth, currents flow in both directions, whereas tides and tsunamis make their way upstream and away from the ocean.

Unlike the riverine current, ocean waves carry not only the fluid, but the state of motion itself, and thus typically travel much faster and penetrate much farther than the fluid particles which they move. Tides penetrate in rivers from several or a few tens of kilometers (in relatively small, steep rivers) to a few hundred kilometers (in large low-land rivers). Estuarine tidal scientists have long been studying propagation of tides into rivers. Over many years, tides have been observed in river estuaries, assigned a number of specific parameters in the process, and analyzed for how these parameters depend on the distance from the river mouth, on the shape of the channel (exponentially narrowing channels are commonly considered), sometimes—on the riverine current, and on some other conditions. A tsunami is just a "short tide," and therefore behavior of well-studied river tides can be extrapolated to tsunamis. Or can it be? Tsunami observations in rivers sometimes display features not known in relation to river tides, nor fully explicable in terms of harmonic constituents employed by tidal hydrodynamics.

This book reflects on recently collected evidence of tsunami penetration in rivers. It will focus on three of the observed phenomena. Firstly, in rivers on the east coast of Japan invaded by the Tohoku tsunami of March 11, 2011, average water levels rose with the tsunami arrival and remained elevated for many hours, even during the tsunami withdrawal phases. Secondly, after crossing the Pacific and losing much of its destructive power, this tsunami propagated into the Columbia River in the USA and clearly showed tsunami's modulation by tide, progressing upriver. Records of earlier tsunamis imply that propagation of relatively small tsunamis in rivers is controlled by tidal phase in a typical manner: receding tide obstructs the intrusion, while high tide facilitates it. And lastly, the 2011 Tohoku tsunami occasioned a set of detailed water level records of a train of bores propagating in a river.

Analysis of these observed phenomena can be a step for improving predictions of the upriver impacts of a tsunami. Modern tsunami forecasts provide, in real time, accurate estimates of when far-traveled tsunami waves will reach a coast, and how high they will be. One of the purposes of this book is to lay a basis for predicting tsunami upriver impacts with the tsunami forecast near the mouth, and a few parameters defining a river's response. Which river parameters quantify tsunami dynamics in a river, how these parameters can be computed knowing a river's morphology, how these parameters can be found using everyday tidal observations—these are some of the questions which this book attempts to answer.

This book employs three methods of investigation, which rely on three different information sources: field data analysis, numerical experiment, and analytical analysis. Field observations, such as water level measurements or the high water marks, are meant to sample ground truth. Thereby wave patterns, which emerge from field observations, presumably convey the ground truth as well. However, finding the physics behind those patterns might be challenging: there are too many factors affecting wave dynamics in a natural river, and there are never enough observations to fully describe the spatial and temporal evolution of the wave. By contrast, numerical experiments supply unlimited observations in a simplified, fully regulated river model with no unknowns. Numerical experiments can greatly aid deducing the wave patterns, but leave the question of whether the simplified numerical river provided adequate representation of real-world conditions. Field data and numerical simulations can be used in the same manner and for the same purpose: to deduce which river conditions, and in which way, affect tsunami intrusion. Differently, an analytical solution provides immediate answers to these questions but often requires further simplification of reality. The three methods aid, guide, verify, and sometimes duplicate each other in our attempts to understand tsunami penetration in a river.

Moreover, the three methods allow the observed wave phenomena to be explained both in layman's terms and in equations of the non-linear shallow-water theory. So, a reader wishing to bypass the extensive use of differential equations can skip analytical chapters. The author tried to compose this book so that the two analytical chapters were complementary rather than mandatory for understanding the rest of the material. This book is not a textbook but can be used as supplemental reading for a course on water gravity waves. The book is intended for specialists in the hydrodynamics of tsunamis and tides, as well as—after omitting the differential equations—for nonspecialized readers interested in natural phenomena.

The author found her determination for writing this book at NorthWest Research Associates, Inc.—a small institute owned and operated by its scientists, brought together by a passion for science and, with occasional luck, by funds to do it. This book grows from a series of four articles published in *Pure and Applied Geophysics* in 2013–2016. Some of the articles are written jointly with Professor Hitoshi Tanaka of Tohoku University, Japan, who has been collecting tsunami observations in rivers, and to whom the author is thankful for sharing these unique datasets. The Preface and Chap. 1 were improved by reviews from Brian Atwater of the U.S. Geological Survey and University of Washington. The author deeply thanks Irina Tolkova and Igor Tolkov for improving English of the original manuscript, and Valeriy Tolkov for troubleshooting all computer-related issues. Water level measurements in rivers in Japan, displayed in this book, came from stations operated by the Ministry of Land, Infrastructure, Transport and Tourism (MLIT), Japan. Gage records in the Columbia River, USA, and at the offshore buoys were provided by the National Oceanic and Atmospheric Administration (NOAA), USA.

Kirkland, WA, USA   Elena Tolkova
October 2017

Contents

1   **Observations of Tsunamis in Rivers**   1
    1.1   How Are Tsunamis Measured?   2
    1.2   Tsunami-Induced Water Accumulation in Rivers   4
    1.3   Tsunami Modulation by Tide: Columbia River   7
    1.4   Tsunami Modulation by Tide: 2010 Chile Tsunami in Rivers
          in Japan   12
    1.5   Bore as a Typical Tsunami Waveform in a River   15
    References   20

2   **First Analytical Chapter: Long Wave Set-Up**   23
    2.1   Cross-River Integrated Shallow-Water Equations   23
        2.1.1   One Difference Between Tide and Tsunami
              Entering a River   27
        2.1.2   Dimensionless Form   28
    2.2   Backwater Curve of a Steady Flow   29
    2.3   Backwater Set by an Intruding Wave   30
        2.3.1   Results to Carry Forward   34
    References   34

3   **Like Tsunamis, Tides Make Rivers Deeper**   37
    3.1   Layman's Physics: A Positive Disturbance Penetrates Farther   37
    3.2   Wave Set-Up in Tidal Observations   40
        3.2.1   Tidal Wave Set-Up in the Yoshida River, Japan   42
        3.2.2   Tidal Wave Set-Up in the Columbia River, USA   45
        3.2.3   Don't Tidal Scientists Know All That?   48
    References   50

4   **Tsunami and Tidal Set-Up in Rivers: A Numerical Study**   51
    4.1   River Models   52
    4.2   Quantifying a River's Response   52
    4.3   Simulation Arrangements and Results   53
    4.4   First Look at the Solutions   54

    4.5   Backwater: Patterns, Relations, Numbers............................. 58
          4.5.1   Backwater Curve.. 61
          4.5.2   Backwater Accumulation Distance 62
          4.5.3   Backwater Height and Wave Period 63
    4.6   Two Metrics for Wave Decay with Distance Travelled................ 63
          4.6.1   Metrics I: Wave Variance...................................... 64
          4.6.2   Metrics II: High Water Marks................................. 67
    4.7   Back to the Real World 69
    References 70

5 **Second Analytical Chapter: Ascending with the Wave** 71
    5.1   Wave-Locked Slope, and a New Form of the Equations 71
    5.2   Unidirectional Wave and Steady Flow 74
    5.3   When Channel Convergence Balances Friction 76
    5.4   Excursus into Analytical Treatment of Estuarine Tides 78
    5.5   Effect of Tidal WLS on a Small Tsunami Ascending a River 81
    References 82

6 **Tsunami Rides Tides** 85
    6.1   Designated Dataset: The 2015 Chile Tsunami in Rivers in Japan 86
    6.2   Metrics for a Tsunami Amplitude and for the Tidal Conditions 89
    6.3   Record's Components: Tide, Tsunami, Background Noise 90
    6.4   Correlation Between an Instant Tsunami Amplitude and WLS 92
    6.5   Methodology of Admittance Computations as a Function of WLS .. 94
    6.6   When Is a Tsunami's Admittance into a Tidal River Greater?........ 98
    6.7   Physics of Admittance Variations with WLS 99
    References 102

7 **Tsunami Bores** 103
    7.1   Shock Equations 103
          7.1.1   Giving Up Momentum Correction Coefficients............... 105
          7.1.2   Shock Conditions in a Channel with Vertical Banks.......... 106
          7.1.3   Shock Conditions in a Channel with Sloping Shores 107
          7.1.4   Whose Flow Is Faster? 109
    7.2   How Does an Ordinary Wave Become a Bore?....................... 110
    7.3   Turbulence or Ripples? 113
    7.4   Shock Reflection from a Barrage: Observations and Analytics....... 113
    7.5   How Good Is This Theory?.. 117
    7.6   The Book's Main Points 119
    References 120

Index.. 121

Chapter 1
Observations of Tsunamis in Rivers

Highlights Tsunami's journey. How are tsunamis measured? NOAA's tide gage network. Japan, MLIT network of water level stations. Super-elevation of the rivers' mean stages during the 2004 Indian Ocean tsunami, the 1983 Japan Sea tsunami, and the 2011 Tohoku tsunami. History of tsunami observations in the Columbia River. Something odd in these records. Tsunami modulation by tidal phase: is it always this way? Spherical focusing on Japan. Chile 2010 tsunami in Honshu rivers. What does a tsunami in a river look like? The 2011 Tohoku-oki tsunami in the Mad River, California. Tsunami bores' journey up and down the Kitakami River, Japan.

A tsunamis starts with a displacement or a push to a water column over an area much greater than the water depth. Typically, tsunamis originate next to the land—at junctures of oceanic and continental tectonic plates. In the Pacific ocean, these tectonic boundaries form the so-called "Ring of Fire", stretched along the oceanic perimeter and showcased by volcanos and earthquakes. Plate movement beneath the seafloor—a submarine earthquake—can result in rapid dislocation of the ocean bottom through several meters over an area tens to hundreds km long and tens km wide. The seafloor deformation transfers to the water column above it. The displaced water—the tsunami source—gives rise to a wave spreading in all directions from the source area. In deep water, the sea surface under the tsunami wave has a tiny slope, while tsunami-induced currents typically do not exceed a few cm/s. However, as the wave approaches a shore and enters shallower waters, it steepens and grows in height. Sea floor topography can funnel the wave and make it even higher and currents stronger at some locations.

Large tsunamis are devastating on nearby coasts. During the 2011 Great East Japan (Tohoku) tsunami, the east coast of an island of Honshu in front of the earthquake experienced waves as high as 20 m,[1] possibly higher in some locations. At the same time, impacts of large tsunamis are not limited to a single region.

---

[1]The wave height about this value was estimated by water marks on coastal cliffs near entrances of Ohtsuchi Bay and Kamaishi Bay in Iwate Prefecture (Mori et al. 2011). A wave height by the coast is not to be confused with a runup height, which reached up to 40 m in this event.

© The Author(s) 2018
E. Tolkova, *Tsunami Propagation in Tidal Rivers*, SpringerBriefs
in Earth Sciences, https://doi.org/10.1007/978-3-319-73287-9_1

Tsunamis propagate across an ocean and can cause wreckage in harbors and flooding even after 10–24 h of trans-oceanic travel. When the wave approaches the shore, its journey ends with running up the beach, or with smashing docks after being trapped in a bay or a harbor, and these are the most prominent faces of a phenomena called "tsunami".

As a tsunami transitions into a river, however, it embarks on another journey, sometimes making many tens of kilometers away from the ocean coast. Traveling along a river, a tsunami intrudes far inland and can cause flooding where it is generally not expected. Great tsunamis of the twenty-first century (such as the 2004 Indian Ocean tsunami and the 2011 Tohoku tsunami) provided plenty of terrifying evidence of the tsunami's exceptional ability to travel long distances up rivers, and inundate low laying areas along the way (Tanaka et al. 2008; Mori et al. 2011; Liu et al. 2013).

1.1   How Are Tsunamis Measured?

> ... When you can measure what you are speaking about, and express it in numbers, you know something about it; but when you cannot measure it, when you cannot express it in numbers, your knowledge is of a meagre and unsatisfactory kind... (Sir William Thomson. Popular Lectures and Addresses, Vol. I, p.80. Macmillan and Co., 1891.)

Once a natural phenomenon becomes a subject of a scientific study, it has to be measured. Tsunamis of the old times are measured by the damage they have made, and by the traces they have left. Damage from a tsunami is expressed by the number of its victims and the amount or cost of houses and infrastructure destroyed. Traces of ancient tsunamis are limited to tsunami deposits—such as sheets of sea sand spreading inland or moved boulders (Bourgeois 2009). Along the Cascadia subduction zone (a 1000-km long tectonic plate boundary that stretches underwater from northern California to Vancouver Island), memories of great local earthquakes and subsequent tsunamis were found mainly in estuaries of tidal rivers[2] (Hutchinson and Clague 2017 and references therein). This geological evidence contributed to a discovery of a previously unknown subduction zone, and to understanding earthquake and tsunami potential in the American Pacific Northwest—a hazard, that had gone unrecognized until the late 1980s.

A recent tsunami leaves more abundant traces, which include sea artifacts deposited inland, water marks on trees, bridges, embankments, and other structures, land stripped of vegetation, moved loads, and knocked down trees. Such traces are seen in photographs in Fig. 1.1, which show the Maule River after the February 27,

---

[2]Evidence of prehistoric Cascadia tsunamis has also been obtained from lakes, e.g. Kelsey et al. (2005). On other coasts, geological histories of tsunamis have been inferred mostly from beach-ridge plains (Monecke et al. 2008; Jankaew et al. 2008; Sawai et al. 2009), or even found in a cave (Rubin et al. 2017).

Fig. 1.1 Traces left by the 2010 Chile tsunami in the Maule River, Chile: a panorama of the river bend at about 7 rkm from the mouth, with views of a scoured bridge (zoomed-in, bottom left) and offshore fishing boats deposited by the tsunami onto a bared riverbank (zoomed-in, bottom right). Post-tsunami survey photographs, courtesy of Hermann M. Fritz, Georgia Tech

2010 Chile tsunami,[3] with views of a scoured bridge and fishing boats picked at the coast and deposited onto a bared riverbank at the river bend, at about 7 km from the river mouth. The height of the water marks helps to deduce the maximal flow depth, while their orientation, severity of scouring, and other actions help to decide on the flow velocity at the time the marks were produced.

However, to capture temporal evolution of the wave, as well as to observe tsunamis too small to leave lasting imprints, one might wish for instruments taking real-time measurements. An instrumented history of Pacific tsunamis starts in 1853–1854, when the US Coast Survey established first self-registering tide gages in San Diego harbor, California, in San Francisco Bay, California, and in Astoria, Oregon, at 29 rkm (river-kilometer, a distance measured along a river) from the Columbia River mouth. Tsunamis followed. As described by Lander et al. (1993), on December 23, 1854, an earthquake with an estimated magnitude of 8.3 off of the southern Honshu triggered a tsunami 21 m high at the local coast. When the tsunami crossed the Pacific and arrived at San Francisco 12 h and 22 min later, and at San

---

[3]Event dates throughout the book are given according to GMT zone, unless a different time zone is specified.

Diego 13 h and 44 min later, it was under 10 cm in amplitude. The next day, another submarine earthquake, estimated as Mw 8.6, happened just 280 km to the southwest of the first one, followed by another tsunami with the largest wave-height of 28 m. By California, this tsunami produced a wave just about 0.5 cm higher than the previous one. The tsunamis were identified there by tidal observers, who noticed their traces in tidal records (Bache 1856; Theberge 2005). A marigram with traces of the first tsunami about 10 h in duration—the earliest confirmed tsunami recordings—can be seen in Lander et al. (1993) and Atwater et al. (2015). Self-operating tide gages made it possible to quantify wave motion in real-time, and to detect tsunamis arriving from far away. Since then, more often than crashing onto a beach, a far-field tsunami is seen as a mere anomaly on a tide gage record.

The Center for Operational Oceanographic Products and Services (COOPS) of the National Oceanic and Atmospheric Administration (NOAA) of the US maintains a tide gage network on the US coasts and in a few large estuaries (which on the West Coast, amounts to the lower Columbia River). Prior to the mid-1990s at most US locations, tidal records were taken as marigrams, which are ink traces on a paper unrolling from a drum with a typical speed of one inch per hour (Lander et al. 1993). Then the paper drums were replaced with digital transmitters, reporting instant water levels every 6 min. To facilitate tsunami observations specifically, most gages started reporting with 1-min intervals since about 2009. All these digital data are freely accessible, and include observations of the 2011 Tohoku tsunami running up the Columbia River.

In Japan, water levels in main rivers are monitored by the Ministry of Land, Infrastructure, Transportation, and Tourism (MLIT) with a network of stations reporting with a 10-min interval. In recent years, some stations started reporting with a 1-min interval. Typically, the stations are separated by several river-km, with the first (most downstream) station located at about 1 km up the river mouth. Rare in the rest of the world and invaluable for our study, the water levels at all gages have a common vertical reference (Tokyo Peil (TP), a national vertical datum near mean sea level of Tokyo Bay). The real-time 10-min sampled data are freely accessible. Starting 2010, a series of tsunamis was recorded in several rivers on the east coast of Honshu.

These recently collected high-quality water level measurements display and quantify three phenomena discussed in this book.

1.2 Tsunami-Induced Water Accumulation in Rivers

The first of these three phenomena is an overall rise in river stage during a tsunami intrusion. This accumulation of water was first inferred from field observations in Sri Lanka, by Hitoshi Tanaka and coworkers. Tanaka et al. (2008) collected and investigated traces left by the 2004 Indian Ocean tsunami in five rivers in Sri Lanka. In particular, the surveyors focused on damages to thirty-four road, railway, and pedestrian bridges in these rivers. By observing the bridges' handrails and guardrails bent seaward, the surveyors concluded that the damage was done by the return tsunami flow, and therefore the return flow velocity was higher than the run-up velocity. Moreover, to leave those marks, the return flow had to occur at a high flow depth. However, on the coast, the return flow coincides with the wave withdrawal and a low flow depth. This discrepancy suggests that, in a river, the water brought by a tsunami had to accumulate before returning to the sea. Tanaka et al. (2008) suggested that water accumulates in a river's floodplain before returning through the river's channel. Therefore, the greater the width of the floodplain relative to the river's width, the greater the return flow. In the Shi Lanka rivers, this assumption can explain the greater damage to bridges across narrow rivers noted by the surveyors. Additional mechanisms of temporal water accumulation in a tsunami event, inherent even in a rectangular channel with a freshwater flow, will be discussed in Chaps. 2–4.

The water accumulation effect appears to spread far upriver—farther than the wave motion can be seen. Tanaka's group interviewed sand/gravel miners who saw the 2004 tsunami ascending the Kalu River. The river's bed slope is very gentle in the downstream area, at 1:5000. The observations were made at several locations along the river, in where the tsunami traveled a long distance. The most inland of the witnesses, 20 rkm upstream from the river mouth, did not observe any form of wave motion, but noticed a slow rise and fall in the water level through as high as 0.6 m (Tanaka et al. 2008).

Abe (1986) describes the same effect in the Agano River on the west coast of Honshu, invaded by a tsunami from the Sea of Japan on 26 May 1983. The tsunami followed a local Mw 7.7 earthquake, and intruded into most of the adjacent rivers. The water level record at 14 km from the Agano River mouth displays mere elevation of the river's surface in response to a tsunami wave train. The upstream river's stage moves up some time after a 2-m high wave crest passes the downstream gage, then slowly relaxes, but never goes below its presumed tsunami-free level (Abe 1986).

The Great Tohoku tsunami of 2011 intruded in every river on the northeast coast of Honshu, and was recorded by all gages which it did not destroy. The tsunami's along-river records at surviving stations present clear evidence for gradual elevation of the river's mean stage, reaching its peak several km upriver (Tolkova et al. 2015) (river's mean stage in this context refers to an average surface elevation during a wave cycle). In particular, the tsunami left records at three locations at 0.5, 4, and 9 rkm in the Naruse River, and at four locations at 0.5, 4.2, 9, and 13.6 rkm in the Yoshida River. Naruse and Yoshida make nearly parallel channels close-by for the first 9 rkm from the coast, separated by a narrow (50–200 m) strip of land. The rivers merge just 0.9 km before entering the Ishinomaki Bay. The first downstream station at 0.5 rkm (Nobiru) is located in the two rivers' common channel. The tsunami was also recorded in the Kobama harbor in Ishinomaki Bay, 11 km from the Naruse/Yoshida common mouth (see maps in Fig. 1.10 for the locations).

These records are displayed in Fig. 1.2. At each in-river station, the river's stage sharply rises by 1–2 m with the tsunami arrival, and remains elevated for many hours. In each river, the second station at 4 rkm recorded a greater elevation of the river's mean stage, than the downstream station. Mean surface uplifts at the

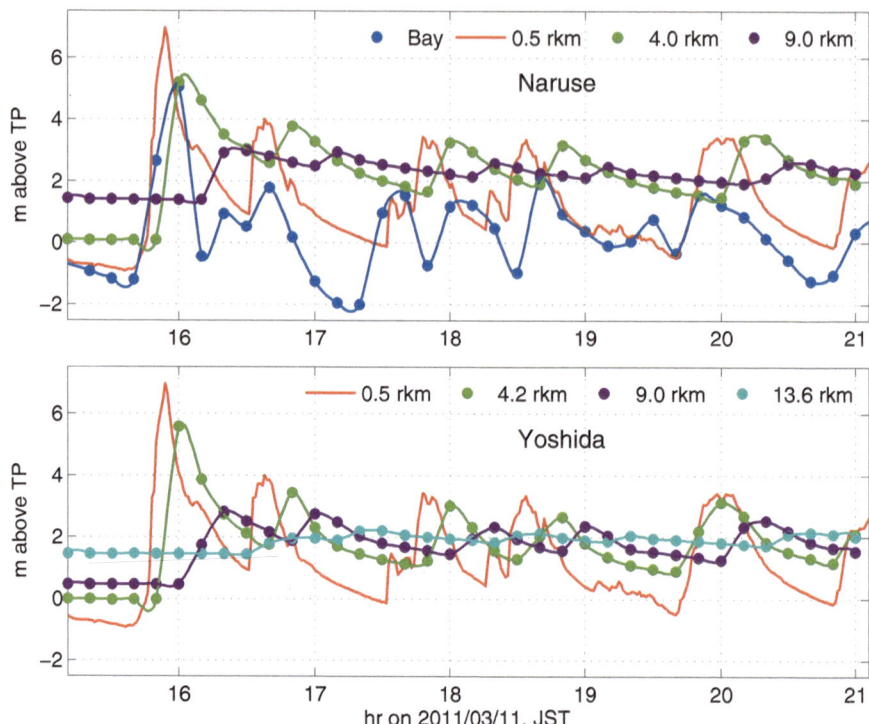

Fig. 1.2 The 2011 tsunami time histories along Naruse (top) and Yoshida (bottom) rivers, at upstream distances (rkm) shown in the plot legends, display an overall rise in river stage during a tsunami intrusion. The water level at Nobiru was recorded with a 1-min interval (dark orange), while measurements at other stations (dots) were taken with a 10-min interval. Additionally, the top panel shows the water level in the Ishinomaki Bay at Kobama harbor (blue). The earthquake happened at 14:46 JST (Japanese Standard Time). Increase in gauge readings due to co-seismic land subsidence at Nobiru was under 0.4 m, decreasing upriver (Adityawan et al. 2014)

more upstream stations (at 9 and 13.6 rkm) exceed the range of remaining wave-like motion. None of this effect is present on the coast, in the Kobama harbor. Kobama's record (Fig. 1.2, top) approximately follows the record at Nobiru, but shows no more rise in the mean water level than can be attributed to the co-seismic land subsidence.

Apparently, during a tsunami intrusion, a river's mean stage super-elevates—a phenomenon common in many rivers. The super-elevation spreads farther upriver than the wave motion penetrates. However, it takes a really large tsunami to cause noticeable water accumulation in a river. Detailed examination of this effect, which might be a key to understanding tsunami penetration in a river, is carried out in Chap. 2 via analytical investigation, and in Chap. 4—via numerical experiments. Moreover, as shown in Chap. 3, tides increase a river's mean depth as well, owing to the same physics. Would not it be possible to scale the river's response for the wave height and period, and then to use everyday tidal observations for predicting a tsunami's behavior in a particular river?

1.3   Tsunami Modulation by Tide: Columbia River

The second of the three phenomena is that tsunami intrusion in rivers is influenced by tidal phase. In the US, this phenomenon was deduced from tsunami observations in the Columbia River (Yeh et al. 2012).

Columbia River is the largest river on the Pacific coast of North America. Its 3.5-km wide mouth is shaped by two powerful jetties, of which the longest southern jetty is about 5.6 km long. Past the mouth, the river forms an estuary up to 12 km wide, scattered with tidal flats, which transitions into a single well-defined channel at Skamokawa at 54 rkm (see Figs. 1.3, 1.4, and 1.5). Tides, almost as high at Skamokawa as they are at the mouth, start diminishing in the fluvial channel, and become barely detectable by the Bonneville Dam at 234 rkm (Yeh et al. 2012; Jay et al. 2015). Tsunamis make their way into the river as well. Lander et al. (1993) display marigrams of four tsunami events recorded at Astoria (29 rkm) on 23 August 1872, 4 November 1952, 23 May 1960, and 28 March 1964. The latter tsunami also reached several upriver stations (Wilson and Torum 1972a). Some of these marigrams, digitized from Lander et al. (1993) and Wilson and Torum (1972a), are reproduced in Figs. 1.6 and 1.7.

Fig. 1.3 View from Astoria, past the 6545 m long Astoria-Megler Bridge, toward the Columbia River mouth. Photo by Bob Heims, U.S. Army Corps of Engineers [Public domain], via Wikimedia Commons. August 1986

Fig. 1.4 View from the ocean coast toward Southern jetty. Columbia River is behind the jetty. Photo by Valeriy Tolkov, June 2014

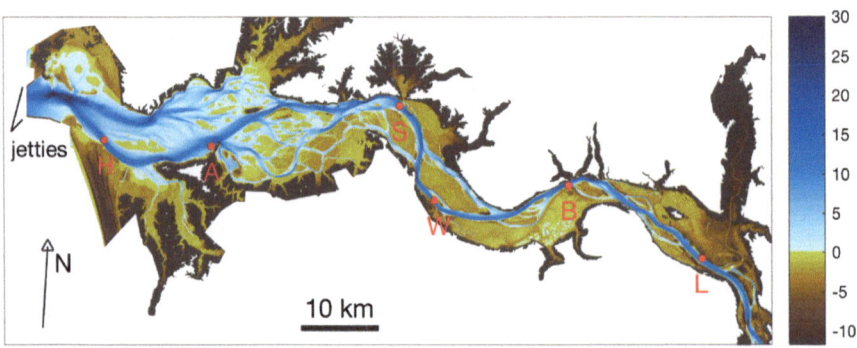

Fig. 1.5 Bathymetric map of the lower Columbia River, a low distortion projection. An arrow points North. Letters refer to gages at Hammond (15 rkm), Astoria (29 rkm), Skamokawa (54 rkm), Wauna (67 rkm), Beaver (86 rkm), and Longview (107 rkm). Colorscale—meters. The illustration uses the Columbia River bathymetry provided for the Workshop on Tsunami Hydrodynamics in a Large River held at the Oregon State University, Corvallis, OR, 15–16 August, 2011 (Yeh et al. 2012)

There Is Something Odd in All These Records

The 11/1952 tsunami came from offshore south-east Kamchatka, caused by an earthquake with magnitude estimates varying from 8.2 to 9.0. About 8 h later, the wave reached the US mainland, arriving almost simultaneously by the Washington and Oregon coasts. The tsunami marigrams at Neah Bay (located in about 10 min wave travel time from the ocean coast, past the entrance of the Strait of Juan de Fuca separating southwest British Columbia, Canada and northwest Washington state, US) and at Astoria (in about 47 min wave travel time up the Columbia River

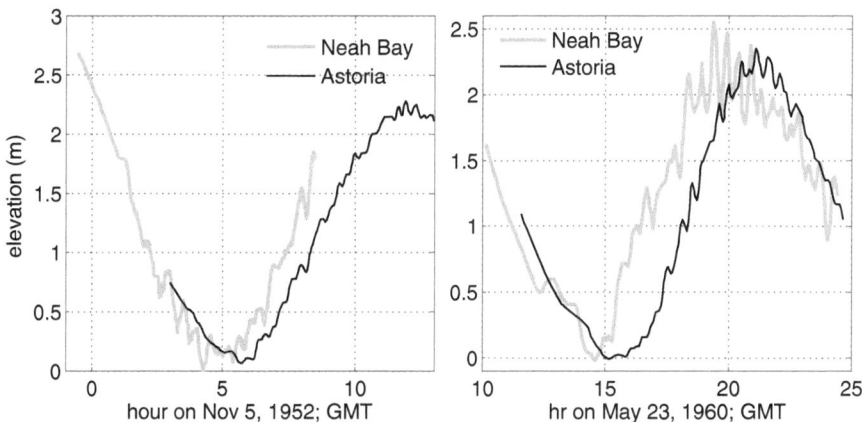

Fig. 1.6 Left: Nov 5, 1952 records at Neah Bay and Astoria tide gages. Right: May 23, 1960 records at Neah Bay and Astoria tide gages. The high-frequency disturbance to what otherwise would be a smooth tidal curve represents the tsunami. Plotted from digitized marigrams. For the original marigrams, see Lander et al. (1993)

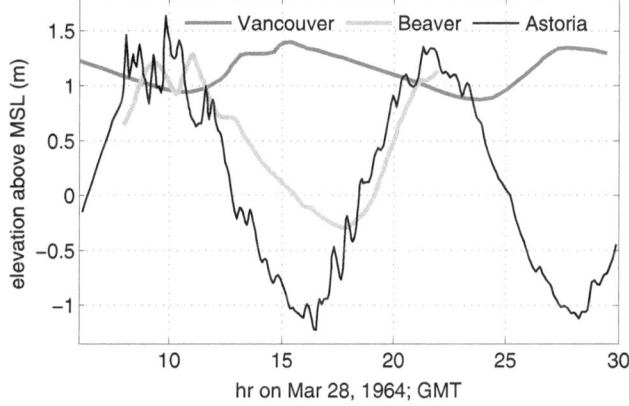

Fig. 1.7 March 28, 1964 records along the Columbia River at Astoria, Beaver, and Vancouver WA tide gages. Plotted from digitized marigrams. For the original marigrams, see Wilson and Torum (1972a)

mouth) are shown in Fig. 1.6, left. Hereafter, the travel time estimates are obtained under the shallow-water approximation, using numerical modeling in an initially still basin filled to the MSL. Based on these estimates, this tsunami should had arrived at Astoria 35–40 min after it came to Neah Bay... but it is barely seen on the marigram until many hours later!

The 05/1960 tsunami came from offshore south-central Chile, generated during the world's largest earthquake with an instrumentally estimated magnitude of 9.5. According to numerical modeling, it takes about 30 min for a tsunami traveling along the continent to span the length of the Washington coast. Given the travel times from the coast to the gages, the tsunami would arrive at Neah Bay and Astoria at about the same time. The respective marigrams are shown in Fig. 1.6, right. The tsunami signal appears sharply on the Neah Bay record, but subdued at Astoria for quite a while.

The 03/1964 tsunami after Mw 9.2 earthquake came from Prince William Sound in central Alaska. It reached Washington coast in slightly under 4 h after the generation. Traveling southbound, it would come to Neah Bay first, and then reach Astoria in under 1 h later. In this case, the tsunami signal appears sharply at both locations, according to the expected timing (Fig. 1.7, see Lander et al. 1993 for Neah Bay). The Great Alaskan tsunami of 1964 propagated into the Columbia River as far as to Portland, OR/Vancouver, WA (two cities on the opposite sides of the river). Its along-river records are shown in Fig. 1.7. The recorded tsunami waves are approximately 0.6 m high crest-to-trough at Astoria, 0.3 m high at Beaver at 86 rkm, and are still detectable at Vancouver, WA at 170 rkm from the ocean.

The next tsunami to enter the Columbia River and to leave clear records at several upriver locations is the Great Tohoku tsunami, generated during a Mw 9.0 earthquake off the East coast of Honshu, Japan, on March 11, 2011. These records at four NOS gages at Astoria, Skamokawa (54 rkm), Wauna (67 rkm), and Longview (107 rkm) (see Fig. 1.5 for the locations), as well as at a DART station 46,404 located in the tsunami path 426 km west of Astoria, are shown in Fig. 1.8. Having an essential duration and high sampling rate (1 or 6 min), these records allow reliable de-tiding. Figure 1.8 also displays the deduced tsunami-only signals at the locations, after the tidal component in the records was removed. Again, the tsunami signal in the river (but not at the offshore DART) is subdued initially, for as long as about 4 h!

These Observations Present Appealing Evidence of Interaction Between Two Long Waves: Tsunami and Tide

Wilson and Torum observed that the tsunami signal in the Beaver and Vancouver records of the 1964 Alaska tsunami can only be detected atop high tide: "Beaver tide gage, in particular, shows that, with the exception of the tsunami waves riding the tide crest, the intermediate waves have lost their identity and hardly register at low tide, though later waves are found again on the succeeding high tide" (Wilson and Torum 1972b). Insights from numerical modeling (Tolkova 2013), and the better quality 2011 data (Yeh et al. 2012) permit to generalize this observation: tsunami dissipation depends on the phase of tide which the tsunami is riding. Tsunami dissipates most with receding tide, dissipates less as the tide starts to rise, and dissipates the least with the high tide.

Indeed, tsunami signals of 1952 and 1960 events at Astoria are low, if visible at all, at their expected arrival times and after, and become strong only when the tide starts to rise. The 1964 tsunami signal appeared at Astoria at the due time, but it arrived with the high tide. Later in the event, the tsunami signal on the rising tide

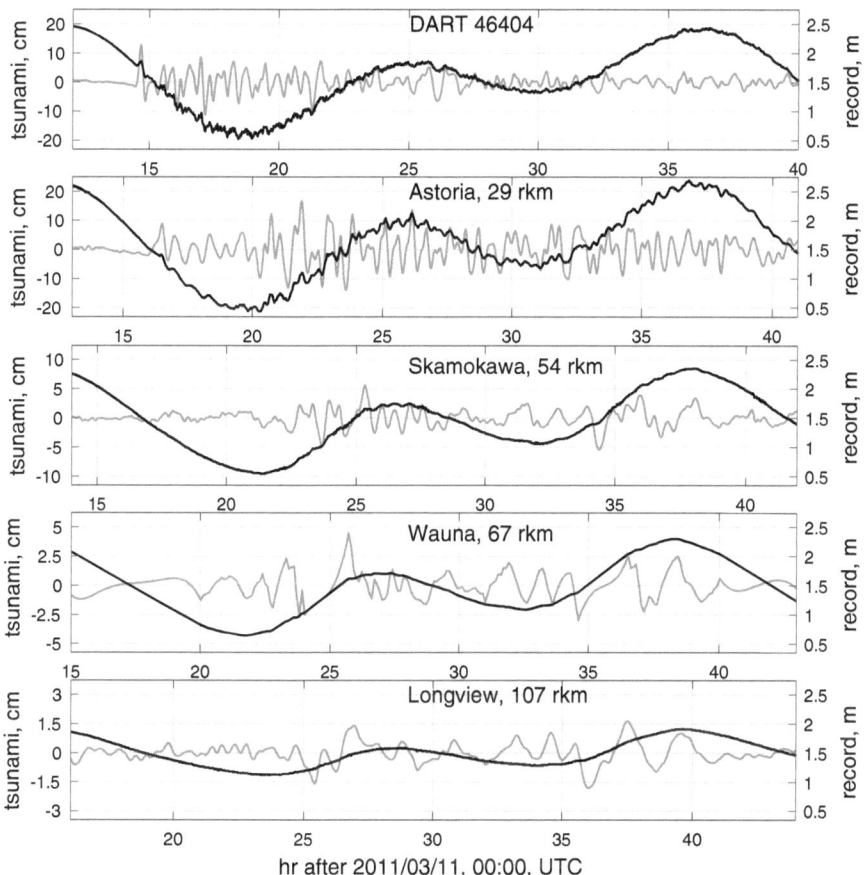

Fig. 1.8 Records of the 2011 Tohoku tsunami offshore Northern Oregon, and along the Columbia River; black—measurements (right y-scale), gray—tsunami signal after de-tiding (left y-scale). The records are sampled with a 1-min interval, except Wauna record, which is sampled at a coarser 6-min interval

seems to be stronger than that on the receding tide. In the 2011 event, the approaching tsunami, as recorded by DART station 46404 offshore Northern Oregon, had higher intensity at the beginning of the wave train (see Fig. 1.8). However, at Astoria, the tsunami signal had low amplitude for the first 4 h, then doubled in intensity on the transition from ebb to flood at 20:00 UTC. As seen in Fig. 1.8, the tsunami riding a receding tide had dissipated entirely by Skamokawa. But with the beginning of the flood tide at 22:00, the tsunami signal at Skamokawa had emerged. At Wauna and Longview, tsunami is not seen above the noise on the transition to the lower low water—neither before the 23-h (Wauna) or 25-h (Longview) mark, nor after the 38-h (Wauna) or 40-h (Longview) mark—but the signal is visible at other tidal phases (Yeh et al. 2012; Tolkova 2013; Tolkova et al. 2015).

What in a tidal river sets the conditions for tsunami intrusion, and generates the observed modulation of the tsunami wave train? Is the pattern of this modulation specific for the Columbia River, or common to rivers in general? Opinions deviated.

1.4 Tsunami Modulation by Tide: 2010 Chile Tsunami in Rivers in Japan

If the entire planet Earth were uniformly covered by a deep ocean, then a wave emanating from a point source would focus on the other side of the Earth, in a point diametrically opposite to the source. If a tsunami originated with an arc-shaped source, it would focus at points where an axis of a circle containing the arc comes out of the Earth surface. In particular, if an arc-shaped source falls on a parallel, the tsunami would focus at the poles. Even with the continents around, and with uneven sea floor topography, this geometrical focusing acts between two major tsunamigenic areas: Japan and Chile. If the North Pole were placed in the Ishinomaki Bay on the east coast of Honshu, then most of the Chilean coast would lay near the 62°S parallel, with relatively obstacle-free ocean in between (see Fig. 1.9).

From Chile to Honshu, a tsunami travels 23 h and covers 5/12 of the Earth circumference. In spite of such a long journey, it keeps some of its power built up by re-focusing. Owing to the MLIT gages and the Earth geometry, many existing tsunami recordings in rivers are those left by Chile-born tsunamis climbing up rivers on the east coast of Honshu.

On early morning of February 27, 2010, at 3:34 local time, a Mw 8.8 submarine earthquake off central Chile generated a trans-Pacific tsunami. The earthquake and the tsunami caused heavy damage and hundreds of fatalities in Chile. Orrego Island, hosting a busy campground in the middle of the Maule River just past its mouth, was completely submerged by a 10-m high wave (Fritz et al. 2011). The tsunami bore swept upriver, and propagated at least 15 rkm—that's where the most upriver eyewitness attempted to shoot a night-time video (H. Fritz, personal communications). The post-tsunami survey group found river banks stripped of vegetation to the bare ground for 10 rkm (which is as far as the surveyors went), where the tsunami deposited offshore fishing boats instead (Fig. 1.1).

Across 5/12 of the Earth circumference, this tsunami again penetrated into rivers, and was observed by forthright witnesses—water level stations. Figure 1.10 shows the water level records of the 2010 Chilean tsunami in the Old Kitakami, Naruse, and Yoshida rivers in Miyagi prefecture, Japan. Old Kitakami is the largest and the least steep of the three, Narise is wider and steeper than Yoshida, and either river is much smaller and steeper than the Columbia River in the US. Naruse and Yoshida merge at 0.9 rkm before entering the Ishinomaki Bay. The first measurement station Nobiru at 0.5 rkm is located in the common Naruse-Yoshida channel.

Measurements of the 2010 Chile tsunami along Naruse, Yoshida, and Old Kitakami clearly show progressive development of tsunami modulation by tide. At

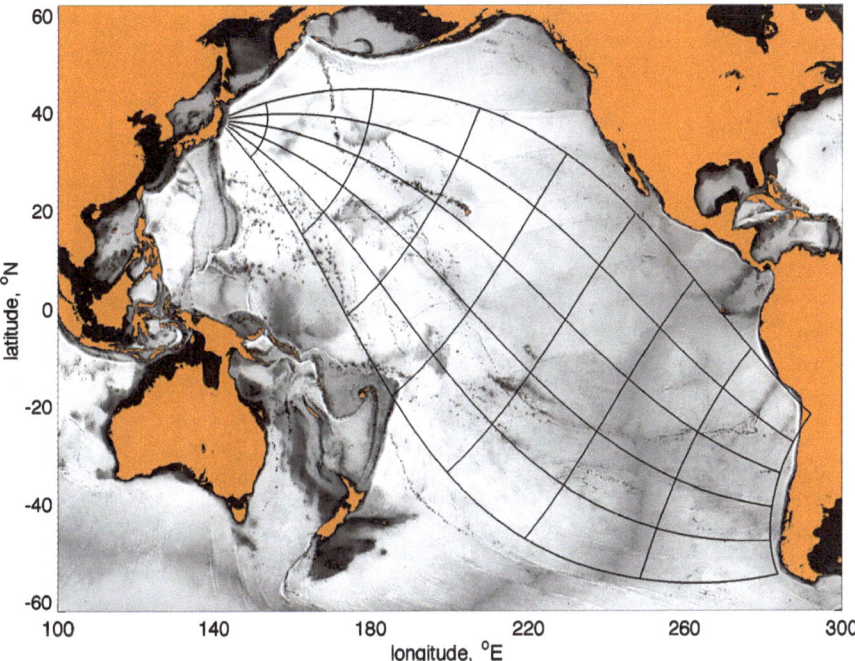

Fig. 1.9 Hypothetical parallels and meridians focusing onto Ishinomaki Bay on the east coast of Honshu. The parallels follow with a 20° interval, starting at 10° around the Bay. The meridians follow with a 14° interval. If the North Pole were in the Ishinomaki Bay, most of the Chilean coast would lay near the 62°S parallel

0.5 rkm in the common Naruse-Yoshida channel, tsunami has uniform intensity over the tidal cycle. However, at 4.2 km up Naruse, the tsunami signal is present only on rising tide, being most intense near the high tide. In Yoshida, the tsunami intruded the greater distance. There, the modulation develops slower than in Naruse. At 4 km upriver, we see the same modulation, but the signal at receding tide still remains. By 9 km upriver, there is no more signal at receding tide; whereas the tsunami is still clearly seen on and near the high tide. Identical behavior is observed in Old Kitakami. In this river, the tsunami propagated the farthest, and was detected as far as at about 33 rkm from the mouth (Tanaka and Tinh 2012). The tsunami modulation by tide in Old Kitakami also develops with the greater propagation distance. No tidal influence on tsunami can be seen near the entrance at 1.24 rkm; but by 22 km up the mouth, the tsunami is fully modulated by tidal phase.

Exactly the same effect has been observed in many rivers: a tsunami reduces to complete disappearance on receding tide, but emerges again on rising tide. At the uppermost station in each river, the tsunami signal totally disappears shortly after the high tide, re-appears right after the lowest tide, and regains its strength on or near the following high tide. An effect so general must have a very basic cause, which we will discuss in Chap. 5 via analytics, and in Chap. 6—with yet another set of water level observations.

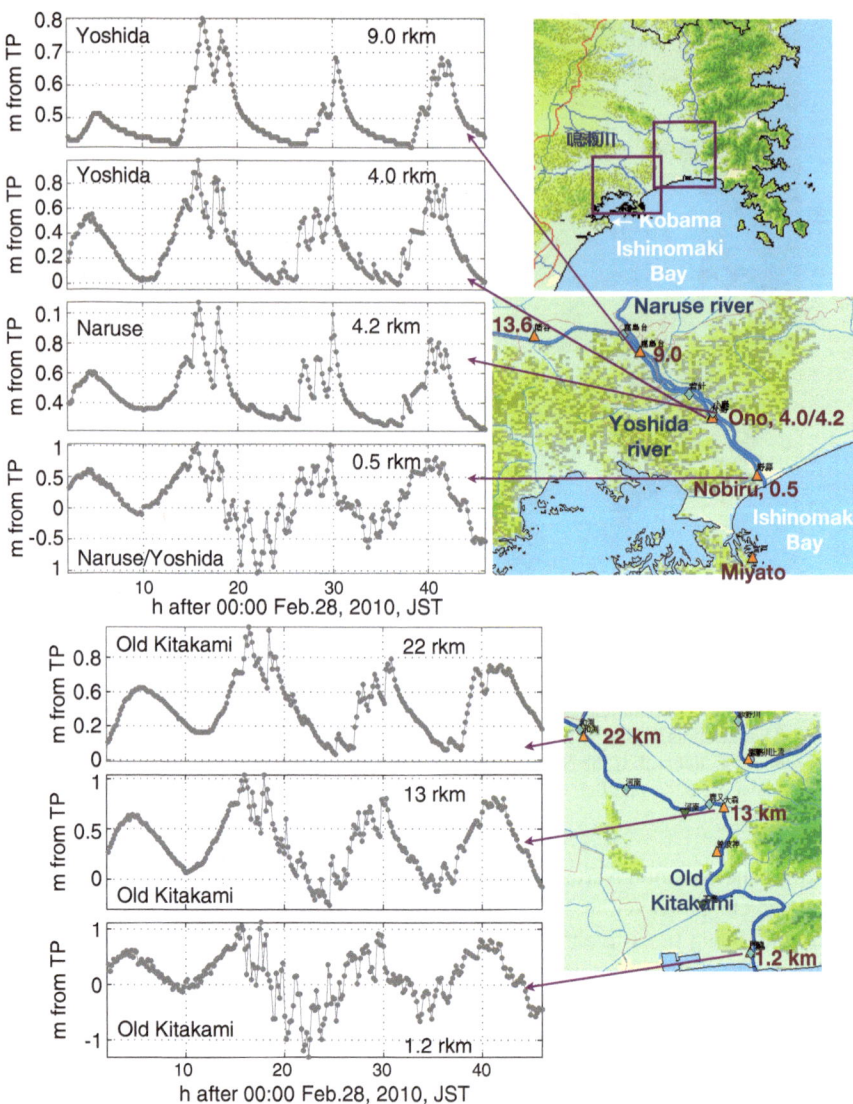

Fig. 1.10 Records of the 2010 Chile tsunami along Naruse and Yoshida and Old Kitakami show development of tsunami modulation by tide. The data are sampled at 10 min intervals. Boxes in the top map delimit areas of the lower maps, Naruse and Yoshida to the west of Old Kitakami. Orange triangles: water level stations. White arrow in the top map points to the Kobama harbor. Maps credit to MLIT

1.5   Bore as a Typical Tsunami Waveform in a River

Not only do tsunamis in a river behave differently than on the coast, but they also look different. The last of the three phenomena discussed in this book is a particular waveform frequently assumed by a tsunami in a river—that of a sudden increase in the water level moving upstream, or a bore.

Ten hours after devastating Japan, the 2011 Tohoku tsunami showed up at the US West Coast. To northern California, it came about 2.5 h before the low tide; and in most locations on the open coast, it was under 1 m hight. The tsunami looked and behaved like a tide speed up by 20–30 times—repeatedly overtopping low beaches, and then withdrawing. A spectator, who came to watch the tsunami on the California coast in Hermosa Beach, remarked that he got bored 10 min later, and started filming the pelicans instead. In bays and harbors set into resonance, the tsunami grew up in height and showed more action, summing up to an over $50-million damage to two dozen Californian harbors (Wilson et al. 2013).

But in rivers, even a low-amplitude tsunami is picturesque. YouTube hosts several eyewitness videos of the 2011 Tohoku tsunami marching up the Mad River in Humboldt county, northern California. Past its mouth, the river flows parallel to the shoreline, separated from the ocean by a low sand spit. According to an eyewitness, four tsunami waves entered the river in a 90 min time frame, varying in height from 1 to 4 ft. The forth, largest wave pictured at 1.5 km from the mouth can be seen in Fig. 1.11. The wave had transformed into a bore—an abrupt jump of the water level, traveling upriver. All four waves captured on videos propagated as bores, but bores of different types. Larger waves, as the one shown in the picture, propagated as bores with turbulent fronts. The small second wave propagated as a so-called undular bore—a bore with a clear smooth front emanating ripples behind it. The third wave, estimated by an eyewitness to be 2–3 ft tall, is initially seen as an undular bore with turbulent edges. Then the turbulence propagates from the edges across the river, closes up, and forms a single turbulent front.

A turbulent bore (or just bore) seems to be a remarkably steady waveform. Sometimes, an undular bore can be seen broken across the river, but a turbulent bore tends to keep straight and consolidated from shore to shore, in spite of channel shape variations or random obstacles to the flow. This appealing waveform, however, had not been captured in measurements—not until a train of bores of the 2011 Tohoku tsunami ascended the Kitakami River in Japan, reflected from a weir at 17.2 rkm, and headed back. This journey was recorded at three locations, and left the most comprehensive evidence to date of bore propagation in a river.

Kitakami River is the largest river on the northeast coast of Honshu, and the fourth largest in Japan. An arial view of the lower Kitakami is shown in Fig. 1.12. The 2011 Tohoku tsunami knocked out a part of a steel 6.5-m high bridge at 4 km from the river's mouth (Fig. 1.13). Low-laying lands along the river were inundated for about 6 km from the mouth. Further upstream, the 5-m high dikes were able to hold the tsunami within the river's flood channel. At 17 rkm, the river is barred by a movable weir, whose gates were closed for the tsunami.

Fig. 1.11 The 2011 Tohoku-oki tsunami hits the Mad River, Arcata/McKinleyville, Northern California (frames from a video record, https://www.youtube.com/watch?v=HYXptqhJq9E&t= 136s). Filmed at 1.5 km up the river's mouth. Top: looking north; bottom: looking south, 14 s later. Courtesy of Jim Campbell-Spickler

The tsunami passage to the weir was recorded by water level stations at Fukuchi at 8.57 rkm, Iino at 14.94 rkm, and the weir at 17.20 rkm (see Fig. 1.12 for the locations). The tsunami records are shown in Fig. 1.14; elevations are counted from TP, and the time is the Japanese Standard Time. The records have a relatively high (for a tide gage) sampling rate of 1 min. Nearly vertical fronts of almost all waves ascending the river imply that the tsunami traveled as a train of bores.

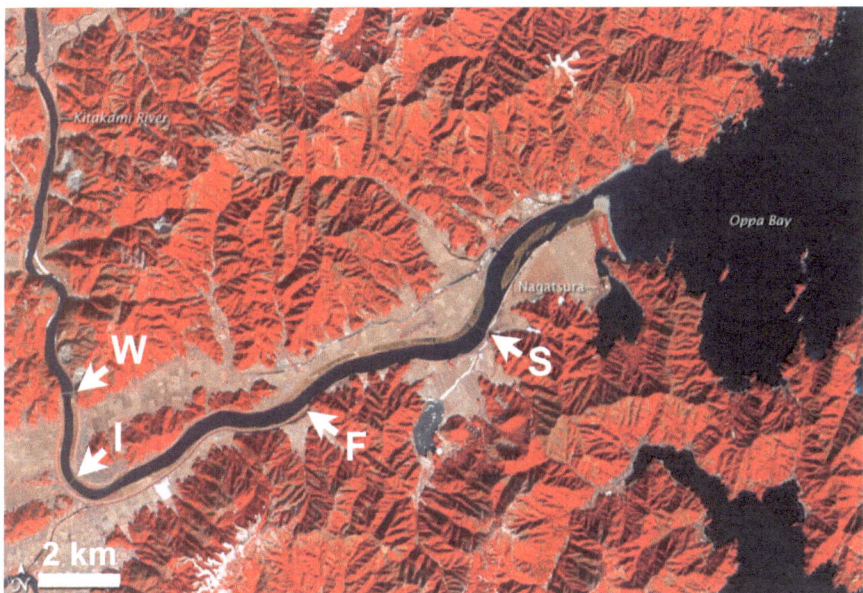

Fig. 1.12 Arial view of the lower Kitakami River on January 16, 2011. Letters refer to observation points: Shin-Kitakami Ohashi Bridge at 4 rkm (**S**); Fukuchi at 8.57 rkm (**F**); Iino bridge at 14.94 rkm (**I**); and the weir at 17.2 rkm (**W**). Image credit to NASA's Earth Observatory: https://earthobservatory.nasa.gov/IOTD/view.php?id=77379

Fig. 1.13 Shin-Kitakami Ohashi Bridge across the Kitakami River at 4 km from the mouth. Its northern sections are washed out by the 2011 Tohoku tsunami. Photo by H. Tanaka

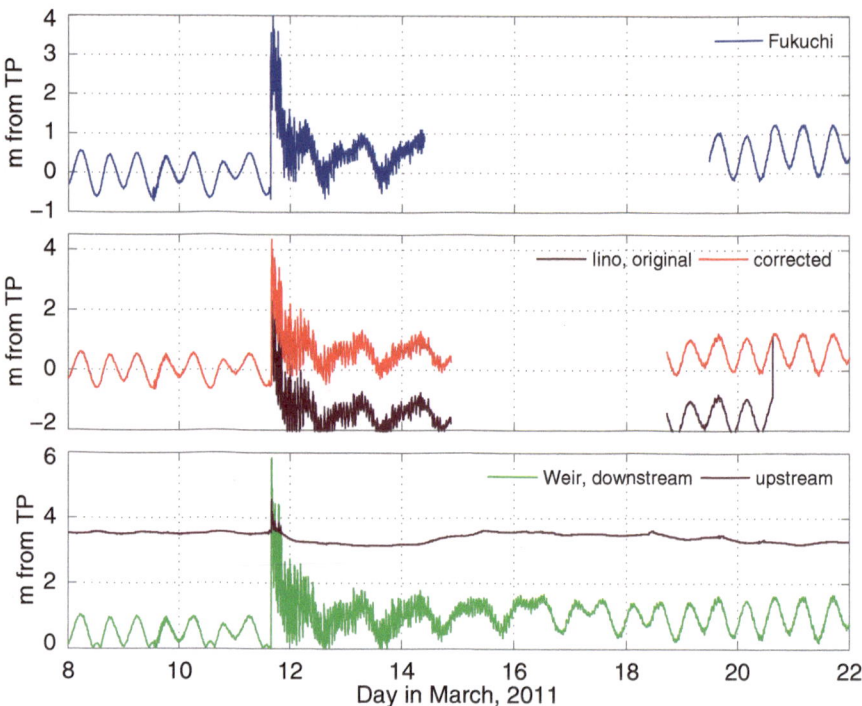

Fig. 1.14 Water level time histories at Fukuchi, Iino, and downstream and upstream the weir. The original Iino record had a vertical shift in the middle segment, starting with the tsunami arrival. Reproduced from Tolkova and Tanaka (2016)

According to the records, the first and tallest tsunami bore measured 4.5-m high at Fukuchi, 3.4-m high at Iino, and created a 6-m high wave by the weir. The weir, rising 3.6 m high above the river surface on the downstream side, was hit with a great force. As seen in the record on the upstream side of the weir, showing slow drop in the water level later in the event, the gates were damaged by the tsunami and started leaking. The greater part of the tsunami was reflected back, while the remaining part overtopped the weir. A 1-m high wave continued upriver. The latter wave reached as far upstream as to a measurement station at 49 rkm, where the river bed elevation was 4.6 m above the sea level (Tanaka et al. 2014). This passage resulted in the longest tsunami intrusion distance in the 2011 Tohoku event. The tsunami eventually set a new river's mean stage elevated by 0.6 m, since the sea level became elevated relative to the land subsided in the earthquake (Adityawan et al. 2014). There is also an apparent super-elevation of the river's mean stage during the main tsunami activity.

Our main focus in these records, however, is the evolution of the tsunami waveform. In Fig. 1.15, top, we see eleven tsunami bores passing by Fukuchi. Approximately 30 min later, each bore comes to the weir, where the same sequence of eleven incident bores superposed with their reflections is clearly seen in the

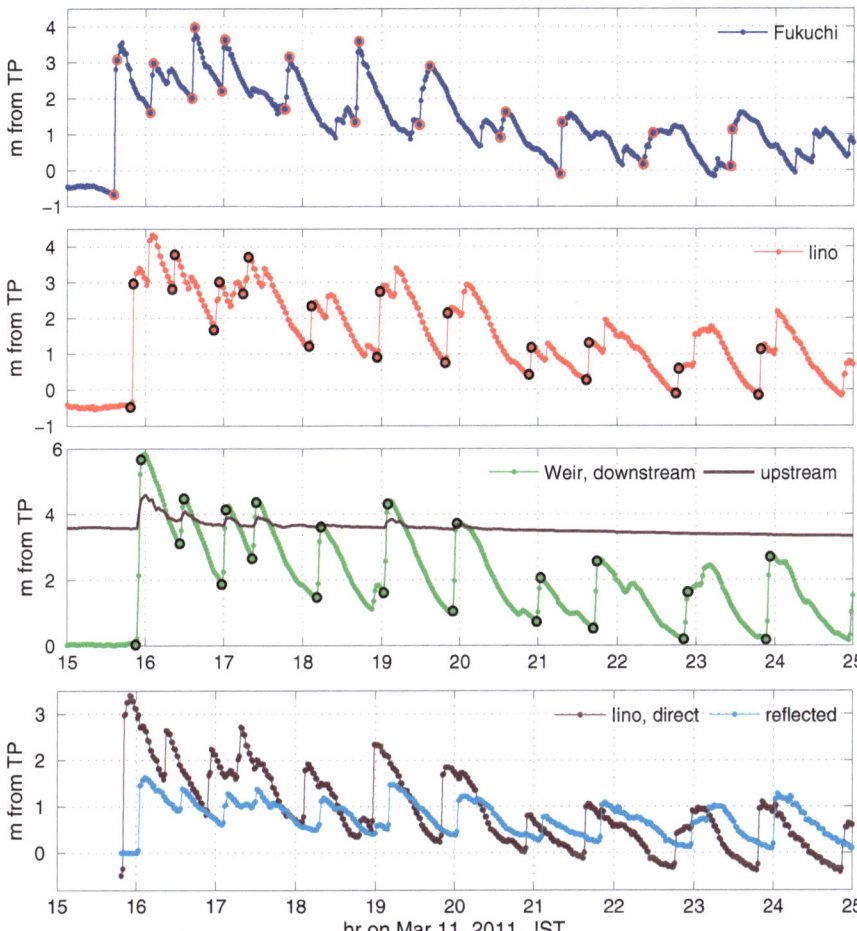

Fig. 1.15 Top three plots: zoomed-in records showing the first eleven tsunami waves at the locations. Circles (red or black) mark readings at the foot and on the top of each upriver-going bore front. Bottom: Iino record decomposed into direct and reflected wave-trains. Reproduced from Tolkova and Tanaka (2016)

record. The circles in the plots delimit each shock front. In Iino record, one can clearly recognize an overlap of incident waves (their fronts are delimited by the circles in the plot) and those reflected from the weir. Tolkova and Tanaka (2016) separated the Iino record into the direct and reflected wave components, shown in the bottom panel in Fig. 1.15. Each component clearly comprises the same train of eleven bores. The wave-train forms are remarkably alike at Fukuchi, in front of the weir, and in the direct and reflected wave trains at Iino. Therefore, each bore kept its shape and identity as it traveled 10.9 km between Fukuchi-Iino-weir-Iino.

Apparently, bores in a channel make a remarkably steady waveform, not affected by (or self-repairing after passing through) the channel's non-uniformities, including those due to significant cross-river inundation. As a result, the bores can be individually traced through the records from Fikuchi to Iino to the weir, and then back to Iino after reflecting from the weir. Most of the reflected bores entirely dissipated before they could return to Fukuchi. In Chap. 7, we adapt the classical one-dimensional shock-wave theory for real-world rivers, and test the adaptation by applying it to infer some parameters of the bores recorded in the Kitakami River.

References

Abe, K. (1986). Tsunami propagation in rivers of the Japanese Islands. *Continental Shelf Research, 5*(6), 655–677.

Adityawan, M. B., Tanaka, H., & Yeh, H. (2014). An estimation of land subsidence due to the 2011 earthquake using measured water level data. *Journal of JSCE, Series B2 (Coastal Engineering), 70*(2), I_216–I_220 (in Japanese).

Atwater, B. F., Musumi-Rokkaku, S., Satake, K., Tsuji, Y., Ueda, K., & Yamaguchi, D. K. (2015). *The orphan tsunami of 1700 – Japanese clues to a parent earthquake in North America* (2nd ed., 135 p.). Seattle: University of Washington Press. U.S. Geological Survey Professional Paper 1707. https://doi.org/10.3133/pp1707.

Bache, A. D. (1856). Notice of earthquake waves on the western coast of the United States, on the 23rd and 25th of December, 1854. *American Journal of Science and Arts, Second Series, 21*, 37–43.

Bourgeois, J. (2009). The geologic effects and records of tsunamis. In E. N. Bernard & A. R. Robinson (Eds.), *The Sea. Tsunamis* (Vol. 15, pp. 55–91). Cambridge, MA: Harvard University Press.

Fritz, H. M., Petroff, C. M., Catalán, P., Cienfuegos, R., Winckler, P., Kalligeris, N., et al. (2011). Field survey of the 27 February 2010 Chile tsunami. *Pure and Applied Geophysics, 168*(11), 1989–2010. https://doi.org/10.1007/s00024-011-0283-5.

Hutchinson, I., & Clague, J. (2017). Were they all giants? Perspectives on late Holocene plate-boundary earthquakes at the northern end of the Cascadia subduction zone. *Quaternary Science Reviews, 169*, 29–49. https://doi.org/10.1016/j.quascirev.2017.05.015.

Jankaew, K., Atwater, B. F., Sawai, Y., Choowong, M., Charoentitirat, T., Martin, M. E., et al. (2008). Medieval forewarning of the 2004 Indian Ocean tsunami in Thailand. *Nature, 455*(7217), 1228–1231. https://doi.org/10.1038/nature07373.

Jay, D. A., Leffler, K., Diefenderfer, H. L., & Borde, A. B. (2015). Tidal-fluvial and estuarine processes in the lower Columbia river: I. Along-channel water level variations, Pacific Ocean to Bonneville dam. *Estuaries and Coasts, 38*, 415–433. https://doi.org/10.1007/s12237-014-9819-0.

Kelsey, H. M., Nelson, A. R., Hemphill-Haley, E., & Witter, R. C. (2005). Tsunami history of an Oregon coastal lake reveals a 4600 yr record of great earthquakes on the Cascadia subduction zone. *Geological Society of America Bulletin, 117*(7–8), 1009–1032. https://doi.org/10.1130/B25452.1.

Lander, J. F., Lockridge, P. A., & Kozuch, M. J. (1993). Tsunamis affecting the West Coast of the United States, 1806–1992 (U.S. Department of Commerce, National Oceanic and Atmospheric Administration, National Environmental Satellite, Data, and Information Service, National Geophysical Data Center, Boulder, CO, USA).

Liu, H., Shimozono, T., Takagawa, T., Okayasu, A., Fritz, H. M., Sato, S., & Tajima, Y. (2013). The 11 March 2011 Tohoku tsunami survey in Rikuzentakata and comparison with historical events. *Pure and Applied Geophysics, 170,* 1033–1046.

Monecke, K., Finger, W., Klarer, D., Kongko, W., McAdoo, B., Moore, A.L., et al. (2008). A 1000-year sediment record of tsunami recurrence in northern Sumatra. *Nature, 455,* 1232–1234. https://doi.org/10.1038/nature07374.

Mori, N., Takahashi, T., Yasuda, T., & Yanagisawa, H. (2011). Survey of 2011 Tohoku earthquake tsunami inundation and run-up. *Geophysical Research Letters, 38,* L00G14. https://doi.org/10.1029/2011GL049210

Rubin, C. M., Horton, B. P., Sieh, K., Pilarczyk, J. E., Daly, P., Ismail, N., et al. (2017). Highly variable recurrence of tsunamis in the 7400 years before the 2004 Indian Ocean tsunami. *Nature Communications, 8,* 16019. https://doi.org/10.1038/ncomms16019

Sawai, Y., Kamataki, T., Shishikura, M., Nasu, H., Okamura, Y., Satake, K., et al. (2009). Aperiodic recurrence of geologically recorded tsunamis during the past 5500 years in eastern Hokkaido, Japan. *Journal of Geophysical Research, 114,* B01319. https://doi.org/10.1029/2007JB005503.

Tanaka, H., Ishino, K., Nawarathna, B., Nakagawa, H., Yano, S., Yasuda, H., et al. (2008). Field investigation of disaster in Sri Lankan rivers caused by the 2004 Indian Ocean Tsunami. *Journal of Hydroscience and Hydraulic Engineering, 26*(1), 91–112.

Tanaka, H., & Tinh, N. X. (2012). The 2010 Chilean and the 2011 Tohoku tsunami waves impact to rivers in the Tohoku Region, Japan. In *Proceedings of 33rd International Conference on Coastal Engineering.*

Tanaka, H., Kayane, K., Adityawan, M. B., Roh, M., & Farid, M. (2014). Study on the relation of river morphology and tsunami propagation in rivers. *Ocean Dynamics, 64*(9), 1319–1332. https://doi.org/10.1007/s10236-014-0749-y.

Theberge, A. E., Jr. 150 years of tides on the western coast: the longest series of tidal observations in the Americas (15 pp.). https://tidesandcurrents.noaa.gov/publications/150_years_of_tides.pdf.

Tolkova, E. (2013). Tide-tsunami interaction in Columbia River, as implied by historical data and numerical simulations. *Pure and Applied Geophysics, 170*(6), 1115–1126. https://doi.org/10.1007/s00024-012-0518-0.

Tolkova, E., & Tanaka, H. (2016). Tsunami bores in Kitakami river. *Pure and Applied Geophysics, 173*(12), 4039–4054. https://doi.org/10.1007/s00024-016-1351-7.

Tolkova, E., Tanaka, H., & Roh, M. (2015). Tsunami observations in rivers from a perspective of tsunami interaction with tide and riverine flow. *Pure and Applied Geophysics, 172*(3–4), 953–968. https://doi.org/10.1007/s00024-014-1017-2.

Wilson, B. W., & Torum, A. (1972a). Runup heights of the major tsunami on North American coasts. In *The Great Alaska earthquake of 1964: Oceanography and coastal engineering* (Committee on the Alaska Earthquake, National Research Council) (pp. 158–180). Washington, D.C.: National Academy of Sciences.

Wilson, B. W., & Torum, A. (1972b). Effects of the tsunamis: An engineering study. In *The great Alaska earthquake of 1964: Oceanography and coastal engineering* (Committee on the Alaska Earthquake) (pp. 361–526). Washington, D.C.: National Academy of Sciences.

Wilson, R. I., Admire, A. R., Borrero, J. C., Dengler, L. A., Legg, M. R., Lynett, P., et al. (2013). Observations and impacts from the 2010 Chilean and 2011 Japanese tsunamis in California (USA). *Pure and Applied Geophysics, 170*(6), 1127–1147. https://doi.org/10.1007/s00024-012-0527-z.

Yeh, H., Tolkova, E., Jay, D., Talke, S., & Fritz, H. (2012). Tsunami hydrodynamics in the Columbia River. *Journal of Disaster Research, 7*(5), 604–608.

Chapter 2
First Analytical Chapter: Long Wave Set-Up

Highlights Cross-river integrated Shallow-Water Equations (SWE). Problem formulation describing a wave ascending a river. One difference between tide and tsunami propagating into a river. Dimensionless form of the SWE, and implications for waves in different rivers. Backwater curve of a steady flow entering a reservoir. Backwater set by an intruding wave. Analytical treatment of the problem: backwater profile set by a small wave ascending a stream.

2.1 Cross-River Integrated Shallow-Water Equations

Long wave propagation in open channels is commonly treated under the 1-D shallow-water approximation in a non-viscous fluid. The one-dimensional approach relies on three premises (in addition to the assumptions behind the shallow-water theory), which are: the channel is straight, though its cross-sectional shape can vary; the surface level is strictly horizontal across the channel; and the flow velocity is uniform in each cross-section. The corresponding system of two partial differential equations, first derived in 1871 by a French mathematician Adhémar Jean Claude Barré de Saint-Venant, expresses the laws of conservation of mass and momentum in application to a slice of fluid contained between two close-by channel cross-sections (Saint-Venant 1871; Stoker 1957). The 1-D equations can also be derived from the more general 2-D shallow-water theory. The latter derivation, given below, accounts for the flow non-uniformity across the channel, as well as keeps track of every simplification made to reduce the dimensionality of the problem.

The Shallow-Water Equations (SWE) in the Cartesian coordinates, expressing mass balance and the balance of momentum in the x-direction, are:

$$h_t + (hu)_x + (hv)_y = 0 \qquad (2.1)$$

$$u_t + uu_x + vu_y + g\eta_x = f \qquad (2.2)$$

where subscripts denote partial derivatives with respect to time t and coordinates x and y in the horizontal plane;

© The Author(s) 2018
E. Tolkova, *Tsunami Propagation in Tidal Rivers*, SpringerBriefs
in Earth Sciences, https://doi.org/10.1007/978-3-319-73287-9_2

h is the flow depth;

u and v are the components of the depth-averaged fluid velocity in the x and y directions, respectively;

η is the free surface elevation above the reference level, so $h = d + \eta$ where d is the river bed elevation counted *down* from the reference level;

g is the acceleration due to gravity; and

f is the friction term, often expressed as $f = -C_d u \sqrt{u^2 + v^2}/h$, where C_d is a dimensionless drag coefficient. In Manning formulation, $C_d = gn^2/h^{1/3}$, where n is the Manning roughness coefficient, typically 0.03–0.04 s $\cdot$m$^{-1/3}$.

The coordinate axes are selected so that the x-axis points upriver with $x = 0$ at the river mouth, and the y-axis is cross-river. The z-axis, when mentioned, is vertical upward.

To obtain the 1-D equations describing the flow as a function of only the upriver coordinate x, the SWE are integrated with respect to y. Two assumptions are due now. Firstly, the river's channel is considered straight (though the results will be extrapolated to natural rivers with their bends). Next, the free surface is presumed strictly horizontal across the river. Consequently, η and η_x are constant within a cross-section, and $\eta_y = 0$. A river's breadth b, a flow area A, a cross-sectionally averaged flow velocity $\bar{u}$, and an average squared velocity $\overline{u^2}$ are defined as:

$$b = y_2 - y_1, \quad A = \int_{y_1}^{y_2} h \cdot dy, \quad \bar{u} = \frac{1}{A} \int_{y_1}^{y_2} u \cdot h \cdot dy, \quad \overline{u^2} = \frac{1}{A} \int_{y_1}^{y_2} u^2 \cdot h \cdot dy, \quad (2.3)$$

where $y_1(x, t)$ and $y_2(x, t)$ are the left and right shoreline coordinates defined by

$$h(y_{1,2}) = 0, \quad or \quad d(y_{1,2}) = -\eta. \tag{2.4}$$

In general, A and b are functions of x and η, determined by the bed shape at a position x along the river and by a local river stage η. Apparently, $A_\eta = b$.

Integrating the continuity equation (2.1) with respect to y and using (2.3) and (2.4) yields

$$A_t + (A\bar{u})_x = 0. \tag{2.5}$$

The momentum equation (2.2) will be multiplied by h and integrated across the river. A few preliminary manipulations making use of (2.1) follow. Firstly,

$$hu_t + hu \cdot u_x + hv \cdot u_y = (hu)_t + u\left((hu)_x + (hv)_y\right) + hu \cdot u_x + hv \cdot u_y \quad (2.6)$$
$$= (hu)_t + (hu^2)_x + (huv)_y$$

Integrating (2.6) across the river and considering (2.3) and (2.4) results in

$$\int_{y_1}^{y_2} \left((hu)_t + (hu^2)_x + (huv)_y\right) dy = (A\bar{u})_t + (m \cdot A\overline{u^2})_x, \tag{2.7}$$

where m is the momentum correction coefficient defined as in (Henderson 1966):

$$m = \bar{u^2}/\bar{u}^2, \quad m \geq 1. \tag{2.8}$$

Secondly, assuming that η_x is constant across the river,

$$\int_{y_1}^{y_2} h \cdot g\eta_x dy = gA\eta_x. \tag{2.9}$$

With (2.7) and (2.9), the cross-river averaged momentum equation becomes:

$$(A\bar{u})_t + (m \cdot A\bar{u}^2)_x + gA\eta_x = A\bar{f}, \tag{2.10}$$

where $\bar{f} = (1/A) \int h \cdot f \cdot dy$.

The unknown coefficient m in the averaged momentum equation is never less then unity. Commonly, m does not deviate far from unity.[1] For instance, consider a flow which occupies the main channel and inundates tidal lands. In the main channel, the fluid moves with velocity u_1, and over tidal lands, it has lost its forward momentum ($u_2 = 0$). The cross-sectional area of the flow is A_1 in the main channel, and A_2 over the inundated land. Then $\bar{u} = u_1 A_1/(A_1 + A_2)$, $\bar{u^2} = u_1^2 A_1/(A_1 + A_2)$, and $m = \bar{u^2}/\bar{u}^2 = 1 + A_2/A_1$. However, typically $A_2 \ll A_1$ due to a low water depth on the inundated land compared to the depth in the main channel, even if the sub-areal widths were comparable (though a large river inundating tidal lands is probably still much wider then the onshore distance of flooding as well). Hence $m \approx 1$. As the third major simplification of this analysis, hereafter $m = 1$.

By setting $m = 1$ and using (2.5), the right part of (2.7) can be further transformed as

$$(A\bar{u})_t + (A\bar{u}^2)_x = \bar{u}_t A - \bar{u}(A\bar{u})_x + A_x\bar{u}^2 + A \cdot 2\bar{u}\bar{u}_x = A(\bar{u}_t + \bar{u}\bar{u}_x) \tag{2.11}$$

Then (2.10) simplifies to:

$$\bar{u}_t + \bar{u}\bar{u}_x + g\eta_x = \bar{f}. \tag{2.12}$$

Next, given that

$$A = \eta b + \int_{y_1}^{y_2} d \cdot dy, \tag{2.13}$$

an average flow depth is defined as

$$\bar{h} = A/b = \eta + \bar{d}, \tag{2.14}$$

---

[1] Tolkova and Tanaka (2016) present a case of a tsunami ascending a river where selecting $m > 1$ was essential.

where

$$\bar{d} = \frac{1}{b} \int_{y_1}^{y_2} d \cdot dy \tag{2.15}$$

is a local sectionally-averaged bed elevation below the reference level. After expressing A and η in (2.5) and (2.12) in terms of $\bar{h}$ and $\bar{d}$, the resulting equation set takes the form:

$$(b\bar{h})_t + (b\bar{h}\bar{u})_x = 0 \tag{2.16}$$

$$\bar{u}_t + \bar{u}\bar{u}_x + g\bar{h}_x = g\bar{d}_x + \bar{f}. \tag{2.17}$$

In a river with steep (or, better, vertical) banks, b and $\bar{d}$ are determined only by the river bed shape and are, therefore, invariant in time. The latter will be our preferred river geometry for carrying out analytical and numerical experiments.

To close the equation set, friction now needs to be expressed in terms of the new state variables $\bar{h}$ and $\bar{u}$:

$$\bar{f} = -\frac{1}{A} \int_{y_1}^{y_2} h \cdot \frac{C_d}{h} \cdot u \sqrt{u^2 + v^2} \cdot dy = -\mu_1 \cdot \mu_2 \cdot m \cdot \frac{C_d}{\bar{h}} \cdot \bar{u}|\bar{u}| \tag{2.18}$$

To arrive at the last expression, the factor C_d/h was taken outside the integral and evaluated as some point with depth h_* in the river, in accordance with the first mean value theorem. In a river with a non-uniform cross-section, the friction is greater on shallow rapids, so typically $h_* < \bar{h}$. Therefore, $C_d(h_*)/h_* = \mu_1 \cdot C_d(\bar{h})/\bar{h}$, anticipating $\mu_1 \geq 1$. The cross-river velocity v under the radical was neglected, and consequently the factor $\mu_2 \geq 1$ was introduced to compensate for the value reduction. After those manipulations, the integral evaluates to $m\bar{u}|\bar{u}|$, according to (2.3) and (2.8). As a result, we adopt a simple expression for the friction term, resembling that for a purely 1-D flow

$$\bar{f} = -\frac{\tilde{C}_d}{\bar{h}} \cdot \bar{u}|\bar{u}| \tag{2.19}$$

where an effective drag coefficient $\tilde{C}_d \geq C_d$ is the product of the true drag C_d and the factors $\mu_1 \cdot \mu_2 \cdot m$ describing the flow non-uniformity. The less uniform the flow across the river, the greater the effective drag. Nonuniform depth across a river, as well as cross-river momentum transfer (e.g., at river bends) contribute toward greater cross-sectionally averaged friction. As a result, computations further in this book may involve "unrealistically" large Manning roughness coefficients.

Hereafter, the bars over variables are omitted, and all qualities refer to cross-river averages, unless stated otherwise.

2.1.1   One Difference Between Tide and Tsunami Entering a River

A wave train ascending a river represents a solution to the SWE (2.16)–(2.17) with the boundary conditions imposed at the mouth and at the upstream end, beyond the wave reach. At the mouth, the surface follows a prescribed motion

$$\eta(t,0) = \phi(t/T), \tag{2.20}$$

where T is the wave characteristic time scale, such as a period for a harmonic wave. At the upstream end, the flow state is defined by a prescribed riverine current:

$$u(t,\infty) = u_\infty. \tag{2.21}$$

For subcritical flows, boundary conditions (2.20)–(2.21) completely determine the motion in a given channel, including the velocity at the mouth and the flow depth upstream (Stoker 1957). In particular, flow velocity varies in co-phase with the surface elevation in a wave entering a still frictionless channel with a flat bottom; whereas the two variables are known to be out-of-phase in a tidal wave propagating into a river. The greater the friction and/or the bottom slope, the greater the phase shift between the current and surface height variations (Proposition A). We will now prove Proposition B, that in the same river, the greater the wave period, the greater the phase shift between these two variables.

   We limit the proof to rivers with a constant bed slope $\beta_0 = -d_x$, and a constant depth and width in the upstream reach. We define Problem I by (2.16)–(2.17) with boundary conditions (2.20)–(2.21). Let $\eta_1(x,t)$ solve Problem I for $T = T_1$, and let $\eta_2(x,t)$ solve Problem I for $T = T_2$. For certainty, $T_1 \ll T_2$. Solution $\eta_1(x,t)$ represents a wave in the tsunami frequency band, and $\eta_2(x,t)$ represents tide, both propagating in a river with a bottom slope β_0, drag coefficient C_d, upstream depth h_∞, and upstream current u_∞. Let γ_1 denote a phase lag between the current and the surface height variations in the first wave (tsunami), and let γ_2 denote the same in the second one (tide).

   By replacing the time and space coordinates t and x with $\tau = t \cdot T_1/T_2$ and $\chi = x \cdot T_1/T_2$, Problem I with $T = T_2$ transforms into Problem II:

$$(bh)_\tau + (buh)_\chi = 0 , \quad u_\tau + uu_\chi + g\eta_\chi = (T_2/T_1)(-\beta_0 + f) \tag{2.22}$$

$$\eta(\tau,0) = \phi(\tau/T_1) , \quad u(\tau,\infty) = u_\infty. \tag{2.23}$$

Let $\eta_3(x,t)$ solve Problem II, and thus describe a wave with a period T_1 in another river carrying the same discharge, but with both the slope and the friction increased by a factor T_2/T_1. According to Proposition A, the phase lag in this solution $\gamma_3 \gg \gamma_1$.

At the same time, Problem II originates with Problem I for $T = T_2$, so their solutions are equivalent:

$$\eta_2(x, t) = \eta_3(\chi, \tau) = \eta_3(T_1/T_2 \cdot x, T_1/T_2 \cdot t) \qquad (2.24)$$

That is, η_2 can be obtained by stretching η_3 through a factor T_2/T_1 along the horizontal axis and in time. The stretching recovers the original river bed shape and the desired wave period, but it preserves the phase relations. Therefore, $\gamma_2 = \gamma_3$, and hence $\gamma_2 \gg \gamma_1$. Proposition B has been proven. We expect a tsunami current to co-oscillate with the free surface when the tsunami freely ascends a river, even though tidal currents can exhibit significant phase lags relative to tidal elevations.

2.1.2 Dimensionless Form

The SWE can be converted to dimensionless variables, indicated below by a 'hat' accent, by selecting a unit of length equal to the river's characteristic depth h_0, a unit of time being $\sqrt{h_0/g}$, and a unit of velocity equal to $\sqrt{gh_0}$. Then all qualities expressing length L, time t, and velocity v in the SWE are replaced with $h_0 \cdot \hat{L}$, $\sqrt{h_0/g} \cdot \hat{t}$, and $\sqrt{gh_0}\hat{v}$, accordingly. The resulting SWE in the dimensionless variables follow:

$$\frac{\partial \hat{h}}{\partial \hat{t}} + \frac{\partial (\hat{h}\hat{v})}{\partial \hat{x}} + \frac{\hat{h}\hat{v}}{\hat{b}}\frac{\partial \hat{b}}{\partial \hat{x}} = 0, \ \ \hat{h} = 1 + \hat{\eta} \qquad (2.25a)$$

$$\frac{\partial \hat{v}}{\partial \hat{t}} + \hat{v} \cdot \frac{\partial \hat{v}}{\partial \hat{x}} + \frac{\partial \hat{\eta}}{\partial \hat{x}} = -C_0 \frac{\hat{v}|\hat{v}|}{\hat{h}^{4/3}}, \ \ C_0 = gn^2/h_0^{1/3} \qquad (2.25b)$$

The river's depth still remains in the equations as a part of C_0 in the Manning friction term. If, however, the wave behavior is known in a river with a particular geometry in a range of friction coefficients C_0, then the wave behavior can be deduced in all rivers with a similar geometry. Moreover, if a certain propagation characterization (e.g., the wave penetration distance) does not depend on C_0 specifically (though it might depend on a related quantity such as the mean surface slope), then this characterization expressed in the dimensionless form is the same in all rivers with a similar geometry. Consequently, in their dimensional form, all such characterizations with a dimension of length are proportional to the river's depth.

2.2 Backwater Curve of a Steady Flow

The steady flow in a river is a stationary solution to (2.16)–(2.17). The continuity equation (2.16) integrates to

$$b \cdot h \cdot u = Q \tag{2.26}$$

where the constant Q is the freshwater discharge rate. Then $u(x) = Q/(b(x)h(x))$, with $Q < 0$ for x-axis pointing upstream, so (2.17) transforms to:

$$h_x \cdot \left(1 - \frac{Q^2}{gh^3b^2}\right) = d_x + \frac{Q^2}{gh^2b^2}\left(\frac{C_d}{h} + \frac{b_x}{b}\right) \tag{2.27}$$

The river stage in the steady flow state can be computed by numerically integrating (2.27), given the bed shape and an initial (which can also be called boundary) condition—a prescribed flow depth at some downstream point, such as at the mouth $h(0)$, or farther offshore. Next, the flow velocity can be obtained from (2.26).

The simplest river morphology is probably that of a river with linearly sloping bed $d(x) = h_0 - \beta_0 x$, in which the flow depth h_0, the bed slope $\beta_0 = -d_x$, the breadth, and the discharge remain constant, and $h(0) = h_0$. Note that, unlike other elevations, the bed elevation is counted downwards. All elevations are counted from the Mean Sea Level (MSL). As follows from (2.17), this flow is possible at a very particular flow speed, at which the friction is balanced by the gravity:

$$gd_x + f = 0, \quad \beta_0 = f/g. \tag{2.28}$$

The river surface elevation above that at the mouth in the uniform flow state is given by $\eta_0(x) = \beta_0 \cdot x$.

In the real world, however, even a river with as simple morphology as above cannot have its depth constant near the mouth. Flow conditions at the mouth depend on geometry of a reservoir, which the river empties into. The river's stage at the mouth is an unknown, which has to be found by integrating (2.27) from some offshore point $x < 0$, where it can be assumed that the surface is indeed at the MSL.

Figure 2.1 shows a numerically simulated[2] surface profile (color-filled) of a hypothetical 1-D river connected to the ocean. The river is represented by a rectilinear channel 5 m deep everywhere (except near the mouth), with 1.5 m/s current, and 0.035 bottom roughness coefficient in Manning friction formulation. The river bed, and the surface in the upstream reach elevate by 32 cm per 1 rkm, which provides the hydraulic gradient compensating the bottom friction for the prescribed discharge. The river is connected to a 1-D ocean which gradually becomes 4500 m deep over an 80 km distance. The ocean surface is at the MSL.

---

[2]All numerical simulations in this book were performed with a tsunami model Cliffs solving the fully-nonlinear SWE (Tolkova 2014; Lynett et al. 2017).

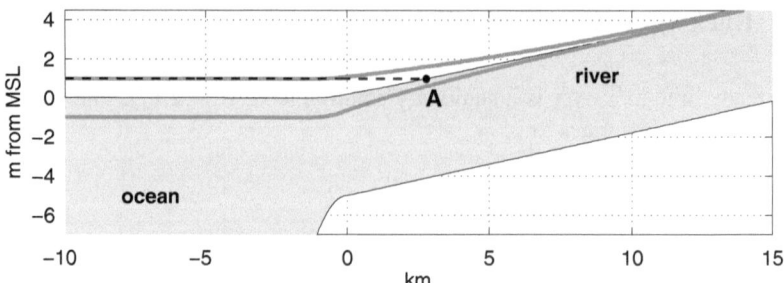

Fig. 2.1 Vertical profile of a river connected to the ocean (color-filled), and two river surface profiles (thick gray) corresponding to the raised by 1 m and lowered by 1 m sea levels (numerical simulations)

The river bed elevation at the mouth is 5 m below the MSL, but the river depth does not stay at 5 m all the way to the sea. If it would, the river surface and the sea surface would meet at an angle. Instead, the river surface elevates above the MSL near the mouth (by 0.18 m at $x = 0$, in this example), and gradually transitions into the sea surface.

The reservoir presents an obstruction to the river flow. The river responds to the obstruction with the so-called backwater effect, which typically refers to a river accumulating its own discharge and super-elevating its surface. The term "backwater" in this context refers to water enclosed between an actual river surface and that corresponding to the hypothetical flow not affected by the obstruction. The river's surface profile upstream of a reservoir is called a backwater curve, which is a transition curve from the river's upstream slope to the horizontal water surface of the reservoir. Backwater effect is felt for tens of kilometers upstream the reservoir, and presents a major concern for dam builders (Southard 2006).

2.3 Backwater Set by an Intruding Wave

A wave entering a river from an ocean also presents an obstruction to the riverine flow, and causes a backwater effect. The river further elevates its mean stage above the static backwater curve, in response to the sea level oscillating about the MSL. In the next chapter, we'll discuss the physics and many manifestations of this type of backwater (also referred to as wave set-up), and its role in setting conditions for wave intrusion in rivers. For now, this phenomenon is to be discovered in the governing equations.

Let us think of a river with a simple geometry ($h = const$, $b = const$) invaded from the mouth by a hypothetical periodic wave. A ready example of such a wave is the tide, if we overlook that tides are not exactly periodic. We are interested in finding a time-average surface elevation ξ_S above the original river stage. The momentum equation (2.17) averaged over the wave period becomes:

$$\overline{u^2}_x/2 + g\overline{\eta}_x = \overline{f}(u, h),\qquad(2.29)$$

where the lines over variables denote the averages. In the absence of the wave forcing, in a river with uniform freshwater flow $u_0 < 0$ and a constant depth h_0, the surface/bed slope is given by

$$g\beta_0 = g\eta_{0,x} = f(u_0, h_0) = f_0.\qquad(2.30)$$

Subtracting (2.30) from (2.29) yields

$$\overline{v^2}_x/2 + u_0 \cdot \overline{v}_x + g\overline{\xi}_x = \overline{f}(v + u_0, h) - f_0,\qquad(2.31)$$

where $\xi = \eta - \eta_0 = h - h_0$ is the surface elevation due wave forcing alone, and $v = u - u_0$ is the wave-induced variation of the flow velocity. Both ξ and v have nonzero means, so we separate a mean and a variation in each wave variable:

$$\xi = \xi_S + \xi_W, \quad \overline{\xi} = \xi_S, \quad \overline{\xi}_W = 0,\qquad(2.32)$$

$$v = v_S + v_W, \quad \overline{v} = v_S, \quad \overline{v}_W = 0.\qquad(2.33)$$

As follows from the continuity equation averaged over the wave period

$$\overline{uh} = \overline{(u_0 + v)(h_0 + \xi)} = u_0 h_0,\qquad(2.34)$$

the mean velocity perturbation is given by

$$v_S = -\overline{v\xi}/h_0 - u_0\xi_S/h_0.\qquad(2.35)$$

As seen from (2.35), the mean current v_S has two oppositely directed components. The first component, directed downstream, describes a return flow compensating the Stokes flux. The second, directed upstream, is a reduction to the freshwater current due to a greater mean flow depth caused by the wave set-up ξ_S.

We will evaluate the wave set-up ξ_S along a river from Eq. (2.31), assuming that the intruding wave has a small amplitude compared to the river's depth: $\epsilon = ||\xi_W||/h_0 \ll 1$. In old traditions of this genre, this assumption is followed by replacing all inconvenient expressions—the friction term in (2.31) in particular— with the power series in ϵ, and truncating the series at some point from where the analytical solution path can be attempted. Note that the phenomenon of interest is another order of magnitude smaller than the wave amplitude at the mouth. As suggested by (2.31), the means v_S and ξ_S are proportional to the square of incoming wave amplitude. Hence it is expected that near the mouth

$$\xi_S \sim \epsilon \cdot ||\xi_W||, \quad v_S \sim \epsilon \cdot ||v_W||, \quad \overline{\xi^2} \approx \overline{\xi_W^2}, \quad \overline{v^2} \approx \overline{v_W^2},\qquad(2.36)$$

and

$$\overline{\xi}/h_0 \sim \epsilon^2 = \overline{\xi^2}/h_0^2. \tag{2.37}$$

However, farther upriver, wave set-up can exceed a range of wave motion. The last equation highlights an important condition for our future manipulations with the power series of a small parameter ϵ: time averaging might level relative magnitudes of subsequent terms in the series, whose magnitudes before averaging were different by a factor ϵ. For example,

$$\xi/h_0 + \xi^2/h_0^2 = O(\epsilon) + O(\epsilon^2), \tag{2.38}$$

but

$$\overline{\xi}/h_0 + \overline{\xi^2}/h_0^2 = O(\epsilon^2) + O(\epsilon^2). \tag{2.39}$$

For further simplification, we assume that the intruding wave does not reverse the flow direction ($v + u_0 \leq 0$), and consequently, the riverine flow is much greater than the wave-induced mean flow perturbation:

$$|u_0| \geq |v|, \quad |u_0| \gg v_S. \tag{2.40}$$

Then friction can be expanded as follows:

$$f = -C_d \frac{u|u|}{h} = f_0 + \frac{C_d}{h_0} \left(\frac{(v + u_0)^2}{1 + \xi/h_0} - u_0^2 \right), \quad f_0 = \frac{C_d u_0^2}{h_0}$$

$$f - f_0 = \frac{C_d}{h_0} [(v + u_0)^2 \left(1 - \xi/h_0 + \xi^2/h_0^2 + O(\epsilon^3) \right) - u_0^2] \tag{2.41}$$

Here, the drag coefficient C_d is assumed to be independent of depth. Considering (2.38) and (2.39), we expand the right part of (2.41), neglecting terms of the third power with respect to wave variables v and ξ. An average over a wave period friction evaluates to:

$$\overline{f} - f_0 = \frac{C_d}{h_0} \left(\overline{v^2} - 2u_0 \frac{\overline{v\xi}}{h_0} + u_0^2 \frac{\overline{\xi^2}}{h_0^2} + 2v_S u_0 - u_0^2 \frac{\xi_S}{h_0} \right) \tag{2.42}$$

Using (2.35) to replace v_S yields:

$$\overline{f} - f_0 = \frac{C_d}{h_0} \left(\overline{v^2} - 4u_0 \frac{\overline{v\xi}}{h_0} + u_0^2 \frac{\overline{\xi^2}}{h_0^2} - 3u_0^2 \frac{\xi_S}{h_0} \right) \tag{2.43}$$

Substituting (2.43) into (2.31), replacing v_S in the left part of (2.31) with (2.35) as well, and regrouping the terms yields the next differential equation for an uplift of the river's mean stage:

$$g \left(1 - \frac{u_0^2}{c^2}\right) \left(\xi_{Sx} + \frac{\xi_S}{L}\right) = \frac{1}{2} \left(\frac{\overline{v^2}}{L_1} - \overline{v^2}_x\right) - \frac{u_0}{h_0} \left(\frac{2\overline{v\xi}}{L_1} - \overline{v\xi}_x\right) + f_0 \frac{\overline{\xi^2}}{h_0^2} \qquad (2.44)$$

with an initial condition $\xi_S(0) = 0$, where

$$L_1 = \frac{u_0^2}{2f_0}, \quad L = \frac{c^2 - u_0^2}{3f_0}, \quad c^2 = gh_0 \qquad (2.45)$$

(using Manning formulation for the friction term results in replacing factor 3 with $10/3$ in the expression for L). The integration is straightforward:

$$g \left(1 - u_0^2/c^2\right) \xi_S(x) = e^{-x/L} \int_0^x F \cdot e^{\chi/L} d\chi, \qquad (2.46)$$

where F denotes the right part of (2.44). The integration can be carried on further using integration by parts, such as:

$$e^{-x/L} \int_0^x \left(\frac{\overline{v^2}}{L_1} - \overline{v^2}_x\right) e^{\chi/L} d\chi$$

$$= \overline{v^2}(0)e^{-x/L} - \overline{v^2}(x) + \left(\frac{1}{L} + \frac{1}{L_1}\right) e^{-x/L} \int_0^x \overline{v^2} e^{\chi/L} d\chi$$

Let us estimate (2.46) assuming that the period-averaged wave intensity decays exponentially upriver with the e-folding dissipation distance L_w:

$$\left(\overline{v^2}, \ \overline{v\xi}, \ \overline{\xi^2}\right) = \left(\overline{v^2}(0), \ \overline{v\xi}(0), \ \overline{\xi^2}(0)\right) \cdot e^{-x/L_w} \qquad (2.47)$$

Note, that conditions (2.47) apply to average squared values, rather than to instant values or their amplitudes, and do not therefore prescribe a particular shape of the oscillation. Integration performed using (2.47) (for $L_w \neq L$) yields:

$$g \left(1 - u_0^2/c^2\right) \xi_S(x) = \frac{e^{-x/L_w} - e^{-x/L}}{1/L - 1/L_w}$$

$$\cdot \left(\left(\frac{1}{L_w} + \frac{1}{L_1}\right) \frac{\overline{v^2}(0)}{2} - \frac{u_0}{h_0} \left(\frac{1}{L_w} + \frac{2}{L_1}\right) \overline{v\xi}(0) + f_0 \frac{\overline{\xi^2}(0)}{h_0^2}\right) \qquad (2.48)$$

The solution for ξ_S is obtained without solving for the wave variables, so it depends on an undefined parameter—the wave dissipation distance L_w. The applicability of the solution is limited to relatively small waves not reversing the flow direction. For example, in a river $h_0 = 5\,\text{m}$ deep ($c = 7\,\text{m/s}$) with $|u_0| = 1\,\text{m/s}$ riverine current, these conditions restrict the maximal wave amplitude at the mouth by $|u_0|h_0/c = 0.7\,\text{m}$.

2.3.1   Results to Carry Forward

The wave set-up profile (2.48) is carried forward as

$$\xi_S(x) = \frac{\xi_0}{\alpha - 1} \cdot \left(e^{-x/L_w} - e^{-\alpha x/L_w} \right), \quad \alpha = \frac{L_w}{L}, \quad L_w = \alpha \cdot \frac{c^2 - u_0^2}{3 f_0}, \tag{2.49}$$

where α and ξ_0 are yet unknown functions of the river's parameters and the incoming wave parameters. As strongly suggested by (2.48), factor ξ_0 is approximately proportional to the wave intensity at the mouth. According to (2.49), the maximal uplift of the river's mean stage occurs at a distance

$$L_\xi = \frac{\ln(\alpha)}{\alpha - 1} \cdot L_w \tag{2.50}$$

and reaches

$$H_\xi = \frac{\xi_0}{\alpha^{\alpha/(\alpha-1)}} \tag{2.51}$$

with the total volume (per unit river width) of the backwater being

$$V_\infty = \int_0^\infty \xi_S(x) dx = \kappa(\alpha) \cdot H_\xi \cdot L_\xi, \quad \kappa(\alpha) = \frac{(\alpha - 1)\alpha^{1/(\alpha-1)}}{\ln(\alpha)} \tag{2.52}$$

The solutions apply in rivers with a simple geometry (uniform width, bed slope, depth, and current in an undisturbed state), and describe effects of relatively small waves not reversing the flow direction.

References

Barré de Saint-Venant, A. J. C. (1871). Théorie du mouvement non permanent des eaux, avec application aux crues des riviéres et a l'introduction de marées dans leurs lits. *Comptes Rendus des Séances de l'Académie des Sciences, 73*(4), 147–154, 237–240.

Henderson, F. M. (1966). *Open channel flow*. MacMillan Series in Civil Engineering (522 pp.). London: Pearson.

Lynett, P., Gately K., Wilson R., Montoya L., Arcas D., Aytore B., et al. (2017). Inter-model analysis of tsunami-induced coastal currents. *Ocean Modelling, 114*, 14–32. ISSN1463-5003. http://dx.doi.org/10.1016/j.ocemod.2017.04.003

Southard, J. (Fall 2006). 12.090 Special Topics: An Introduction to Fluid Motions, Sediment Transport, and Current-Generated Sedimentary Structures. MIT OpenCourseWare: Massachusetts Institute of Technology.

Stoker, J. J. (1957). *Water waves*. New York, NY: Interscience Publishers Inc.

Tolkova, E. (2014). Land-water boundary treatment for a tsunami model with dimensional splitting. *Pure and Applied Geophysics, 171*(9), 2289–2314. https://doi.org/10.1007/s00024-014-0825-8.

Tolkova, E., & Tanaka, H. (2016). Tsunami bores in Kitakami river. *Pure and Applied Geophysics, 173*(12), 4039–4054. https://doi.org/10.1007/s00024-016-1351-7.

Chapter 3
Like Tsunamis, Tides Make Rivers Deeper

Highlights A river meeting an ocean - How far upriver is tidal motion felt? Layman's physics: a positive disturbance penetrates farther. Two plus one explanations for the backwater due to wave forcing. River stage rise/fall without wave motion in upriver areas owing to a tsunami intrusion or the spring-neap tidal cycle. Approach to calculating the tidal set-up height. Tidal wave set-up in Yoshida (Japan) and Columbia (USA) rivers. More than a harmonic constituent.

3.1 Layman's Physics: A Positive Disturbance Penetrates Farther

Should the sea level rise, a river would respond with elevating its stage for tens of kilometers upstream from the mouth, even where its stage was already higher than the new sea level, as shown in Fig. 2.1 of the previous chapter. This is called the backwater effect, well-known in hydraulics of open channels. A wave entering a river from the ocean also presents an obstruction to the riverine flow and causes a backwater effect. The river further elevates its mean stage above the static backwater curve, in response to the sea level oscillating about the MSL. However, the backwater profile due to sea level oscillations is very different from the backwater curve next to a still reservoir. The height of backwater hold back by a reservoir is maximal at the mouth and gradually diminishes upriver. In contrast, the height of backwater hold back by an intruding wave is negligible at the mouth—where the surface level is prescribed by the sea level—and reaches its peak at some distance upriver. This second kind of backwater (also called "wave set-up") is very pronounced in the 2011 tsunami observations along several rivers in Japan; but it is also present at any time in every river emptying into a sea, which invades the river with tides. Tides make a river deeper than it would be if the ocean were still. Tides increase the river's depth far beyond the downstream reach where the tidal variations can be seen. How far upriver do tidal effects penetrate? Imagine a periodic wave propagating into an endless frictional channel with a flat bottom ($d_x = 0$) and no outgoing current. The backwater effect in such channel if felt indefinitely, that is, $\eta(\infty) = \eta^{setup} > 0$ due

© The Author(s) 2018
E. Tolkova, *Tsunami Propagation in Tidal Rivers*, SpringerBriefs
in Earth Sciences, https://doi.org/10.1007/978-3-319-73287-9_3

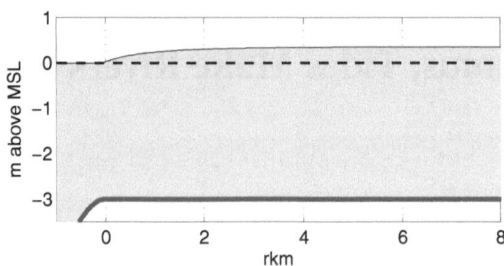

Fig. 3.1 Schematic time-averaged vertical ($x - z$) profile in a frictional channel invaded from the mouth by a periodic wave: water (color-filled), bottom (thick gray), the original surface level (dashed black). In the established regime, the channel's mean surface elevates starting from the mouth, and remains elevated indefinitely

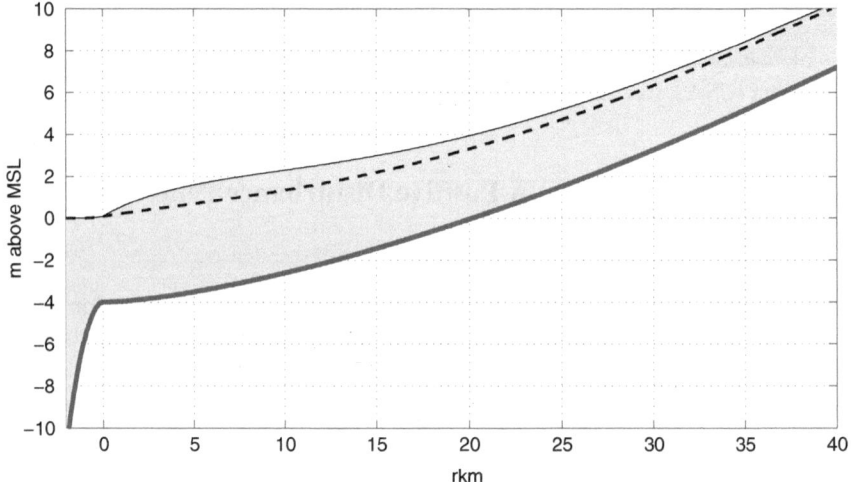

Fig. 3.2 Schematic time-averaged vertical ($x - z$) profile in a hypothetical river invaded from the mouth by a periodic wave: water (color-filled), bottom (thick gray), the original river stage (dashed black). Wave-induced backwater, or wave set-up, refers to the water accumulated above the original river stage

to the wave setup (Fig. 3.1). In a river with a sloping bottom and freshwater flow, the backwater is localized in a finite downstream segment (Fig. 3.2).

The higher amplitude of the intruding wave, the higher the wave set-up. Therefore, sub-tidal (defined as having periods significantly longer than 1 day) water levels along the river co-oscillate with the tidal range at its mouth. These sub-tidal oscillations "may persist further landward, even when the D1 and D2 constituents of the tide are no longer observable" (Hoitink and Jay 2016)—in agreement with tsunami observations at the uppermost point, where "no form of wave motion could be observed, only a slow rise and fall in the water level" (Tanaka et al. 2008). Hoitink and Jay (2016) report that sub-tidal oscillations of the river stage can be detected as fast as 1000 rkm from the coast in the Amazon river, 730 rkm—in Saint Lawrence,

600 rkm—in Yangtze, and 234 rkm—in the Columbia river, where the ocean influence is cut off by a physical barrier (Bonneville Dam). Buschman et al. (2010) pointed out that different tidal wave set-ups in connected channels comprising a river delta affect the flow division and may even reverse the flow circulation.

That oscillatory sea surface motion holds back some more water in a river can be justified in more than one way. Firstly, it can be deduced from the backwater effect of the first kind. For a wave with a very long period, the river's response to the sea surface motion can be approximated by stationary backwater curves. This means that the river continuously adjusts to the changing sea level while keeping the river's discharge constant at every cross-section, at all times. Figure 2.1 of the previous chapter shows backwater curves in a hypothetical 1-D river, for three sea levels: undisturbed (MSL), and shifted 1 m up and 1 m down from MSL. As seen in the figure, a sea rise of 1 m inundates the original river surface for only 3 rkm inland (to point A), but elevates the actual river's stage for more than 10 rkm. On the other hand, a 1 m deep trough lowers the river surface over the shorter distance of about 5 rkm. Since the positive phase penetrates farther upriver, the average of the two curves yields an uplifted stage. In Sect. 4.5, this reasoning will help to find the lower limit for the backwater profiles created by waves with different periods.

Apparently, in the above case of the raised sea level, the sea did not penetrate past point A, but the river itself filled in most of the space under the elevated river stage, as well as resisted the stage loss during sea withdrawal. This wave set-up mechanism highlights the riverine flow as an instrument to propagate a positive (elevated) disturbance from the ocean and to dampen a negative one. Consequently, it does not apply in a flat channel with no outgoing current.

Secondly, the wave set-up can be explained by friction upon the wave-induced currents. The greater the flow depth and lower the velocity, the smaller the frictional losses in the intruding wave. Since inflow occurs at a greater flow depth and, in the presence of riverine flow, at lower velocities than drawdown, frictional resistance to inflow is smaller than to drawdown. Again, a positive disturbance (wave peak) propagating upriver experiences less attenuation than a negative one (wave trough) roughly coinciding with outflow. Thereby at some distance from the mouth, the positive disturbance prevails and thus elevates the river's mean stage. This wave set-up mechanism applies both in a river and in a still frictional channel.

The two mechanisms add up to make the river's response asymmetric toward positive and negative phases of the intruding wave: a positive wave phase penetrates further upriver than a negative one. Consequently, wave troughs get "erased"—they become flattened and therefore widen, whereas wave peaks remain sharp, yielding typical upriver waveforms seen in tidal and tsunami records. As a result, the river responds to sea level oscillations with the uplift of the river's mean stage, which reaches its peak at some distance from the mouth.

Finally, the third interpretation of the wave set-up in a river is the one generally accepted in tidal literature. This interpretation follows the work of LeBlond (1979), who balanced sub-tidal variations of friction and of the surface elevation gradient. Since then, backwater accumulation is explained by steepening of the river's surface slope. The greater surface slope supplies the hydraulic gradient needed to (1) release

the river's discharge against friction increased by tides and (2) release an additional
return discharge compensating the Stokes flux[1] (Jay and Flinchem 1997; Buschman
et al. 2010; Sassi and Hoitink 2013; Jay et al. 2015; Hoitink and Jay 2016).

3.2 Wave Set-Up in Tidal Observations

Tsunami has a big sister—tide. Even though a typical tsunami's wavelength is
an order of magnitude shorter than that of the tide, both tides and tsunamis are
considered long waves which move the entire water column enclosed between
the sea floor and sea surface. The two waves have different origins: tsunamis are
triggered by initial displacement of water within a relatively small area, while tides
are generated by variations in the Moon's and Sun's gravitational attraction all over
the ocean. In a river, however, both types of waves are forced in the same manner—
by water motion at the river's mouth. Therefore mathematically, tide's and tsunami's
propagation in rivers is governed by the same shallow-water equations. Tides are
ever-present. Tsunamis do not happen often. Great tsunamis, like the 2011 Tohoku
tsunami, happen in the same location about once in 300–500 years—a worrisome
detail for the US Pacific Northwest where such a tsunami had last occurred in 1700
(Atwater et al. 2015). Can tidal observations in rivers on the US West Coast be used
for predicting behavior of the next large tsunami in there?

As suggested by the 2011 tsunami observations, a tsunami intruding into a river
can cause water accumulation (wave set-up) in affected reach. Rivers flowing into
oceans are constantly invaded by ocean tides. By how much do tides elevate the
river's mean stage? One cannot switch the tides off and observe an undisturbed
position of the river surface. However, one can observe the river's response at
different phases of the synodic cycle,[2] track whether spring tides cause the stage to
vary about a higher sub-tidal mean level than neap tides do, and attempt to evaluate
the river's stage as a function of the height of the intruding wave. Hereafter, only
diurnal, semi-diurnal, and shorter-period tidal constituents will be referred to as
Tides, whereas surface level variations on longer time scales will be referred to as
variations of the Sub-Tidal Mean Sea Level (ST-MSL) $s(t)$ or Sub-Tidal Mean
River Stage (ST-MRS) $r(x, t)$. Specifically, a 2-day (48 h) cut-off period is selected
to separate Tides and sub-tidal motion. ST-MSL variations penetrate in the river
and become ST-MRS variations. It is reasonable to expect that a contribution of ST-

---

[1]The Stokes flux refers to a positive (in the direction of the wave propagation) mean discharge at
any fixed location x generated by a simple harmonic wave, with zero mean surface displacement.
Since mass transport in the positive direction occurs at the higher flow depth than that in the
reverse direction, the mean discharge over a wave period is positive, which cannot take place in an
established situation when a system is to periodically recover its state. Thereby the system must
compensate for the Stokes flux by developing a mean return flow.

[2]An approximately 15-day-long cycle (half of the synodic month) during which a tidal range
becomes larger than the average (spring tides) and lower than the average (neap tides).

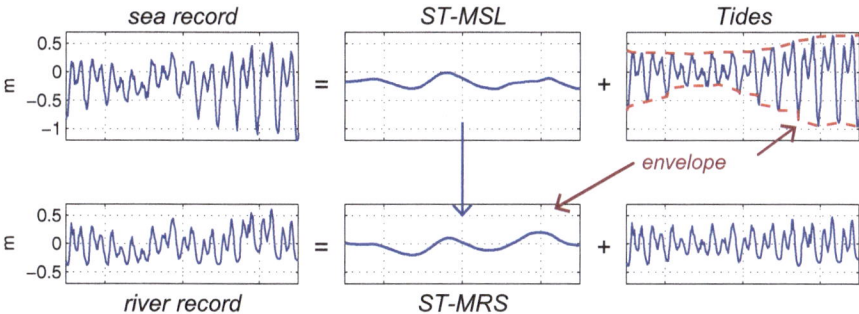

Fig. 3.3 Water level records separated into sub-tidal MSL or MRS, and Tides. Variations of both ST-MSL and tidal envelop height contribute to ST-MRS

Fig. 3.4 Schematic sub-tidal $x - z$ profile in a river: water (color-filled), bottom (thick gray), ST-MSL elevated by s, ST-MRS elevated by r, static backwater curve at elevation b corresponding to the ST-MSL (dashed black), and tidal wave set-up height w

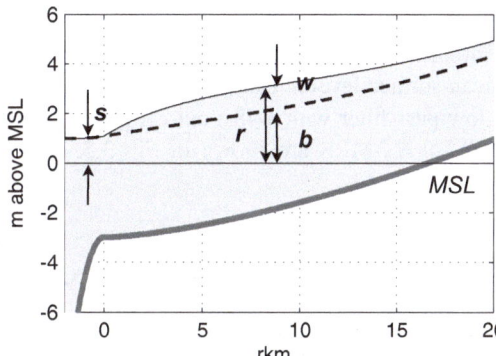

MSL into ST-MRS is given by the backwater curve continuously adjusting to the changing sea level. The elevation of the backwater curve $b(x, s)$ is approximated by a shifted and scaled copy of the ST-MSL s:

$$b(x, s) = \beta_0(x) + \beta_1(x)s, \tag{3.1}$$

where β_1 gives the fraction of the ST-MSL variations present in ST-MRS. Unless a river has morphological peculiarities, $\beta_1(x)$ diminishes upriver starting from $\beta_1(0) = 1$ at the mouth. Tides also elevate ST-MRS through a value w dependent on the strength of tidal forcing (Fig. 3.3). The latter strength will be quantified here simply by the height of tidal envelope at the river mouth $p(t)$, estimated as a range of Tides within a sliding 25-h window centered around t (a more sophisticated metrics expressing an instant wave amplitude is used in Chap. 6). Then our simple model of the sub-tidal river's stage at point x is composed of the height of the backwater in respond to ST-MSL, the wave set-up in respond to Tides, and an occasional contribution $\epsilon(x, t)$ of the upstream-end factors affecting the river stage, such as variations of the freshwater discharge (Fig. 3.4):

$$r(x, t) = b(x, s) + w(x, p) + \epsilon(x, t). \tag{3.2}$$

Next, we demonstrate that actual water level records indeed decompose according to (3.2), yielding small (most of the time) residuals ϵ; and determine the function $w(x, p)$.

3.2.1 Tidal Wave Set-Up in the Yoshida River, Japan

An average bed elevation in the Yoshida River is about 5 m below TP (Tokyo Peil) at the mouth, and 2 m below TP at Ono, with an average bed slope of 23 cm per 1 rkm for the first 15 rkm. Tides penetrate into Yoshida for about 15 rkm, and cover this distance in under an hour. Here, we examine hourly water level records provided by MLIT in the Yoshida River at Nobiru at 0.5 rkm, Ono at 4 rkm, Kashimadai at 9 rkm, and Miyato in the Ishinomaki Bay, near the river mouth (see Fig. 1.10 for the locations). The records span the winter of 2014–2015 from mid-October to early March, a time frame chosen for milder variations of river discharge. A sub-tidal mean surface level at each location is extracted from a corresponding record using a low-pass filter with a 48 h cut-off period. The records and the corresponding ST-MSL or ST-MRS are shown in Fig. 3.5, and zoomed-in in the lower panel. The

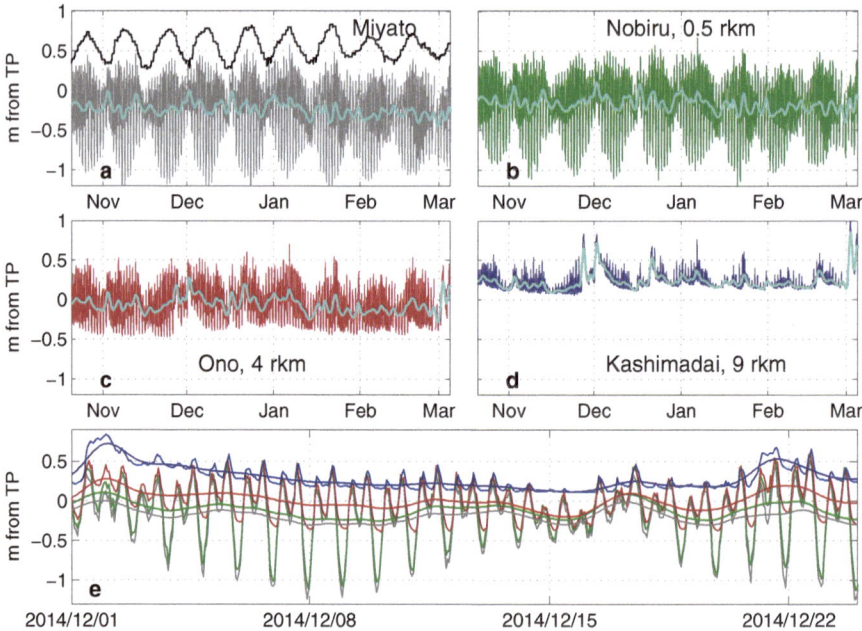

Fig. 3.5 (**a**) A water level record at Miyato s, half-height of tidal envelope p (black), and ST-MSL (cyan); (**b**)–(**d**) water level records along the river and ST-MRS r in each record (cyan); (**e**) zoom-in into overlapped tidal records at Miyato (gray), Nobiru (green), Ono (red), and Kashimadai (blue), and ST-MRS/ST-MSL in each record

record at Nobiru closely follows the coastal record, and ST-MRS there closely follows ST-MSL. The record at Ono shows excessive attenuation at wave throughs, and the elevation of ST-MRS curve above ST-MSL varies. Only tidal crests, greatly reduced in height, penetrate to Kashimadai.

To remove contribution of ST-MSL variations into ST-MRS, sub-tidal MRSs $r(x_j, t)$, where j refers to either Nobiru, Ono, or Kashimadai, are fitted with ST-MSL $s(t)$ and a constant function. Fit with the constant function yields a residual with a zero mean over the whole record. The upper panel in Fig. 3.6 shows the height of the tidal envelope at Miyato $p(t)$ (scaled down five times for plotting), and the fit residuals at the three stations

$$\hat{r}(x_j, t) = r(x_j, t) - \big(\beta_0(x_j) + \beta_1(x_j)s(t)\big) = w(x_j, p(t)) + \epsilon_j - a_j, \qquad (3.3)$$

where a_j is a constant selected to set $< \hat{r} > = 0$; $< \cdots >$ denote a record's mean. Fit coefficients show which fraction of MS-MSL variations is felt at the gage location: $\beta_1 = 1.0$ at Nobiru, $\beta_1 = 0.9$ at Ono, and $\beta_1 = 0.5$ at Kashimadai. As seen in the plots, the variance in the residual $\hat{r}(t)$ at Nobiru is low, which means that ST-MRS near the river mouth is almost entirely prescribed by ST-MSL (at least, outside extreme weather events) with little or no contribution from the wave set-up. At both Ono and Kashimadai, however, the residual $\hat{r}(t)$ is significant and dominated by the

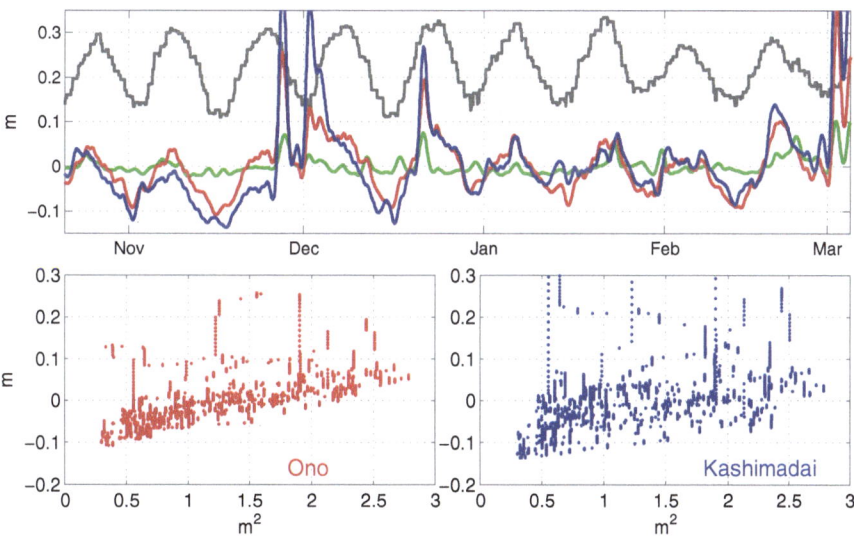

Fig. 3.6 Top: ST-MRS with the contribution from ST-MSL variations removed (residual $\hat{r}$) at Nobiru (green), Ono (red), and Kashimadai (blue), and 1/5 height of the tidal envelope at Miyato p (gray). The residual $\hat{r}(t)$ at Nobiru (green) is low and shows little correlation with the envelope of incoming tide (gray), whereas the residuals at Ono (red) and Kashimadai (blue) are significant and co-vary with the tidal envelope. Bottom: the same residuals $\hat{r}$ at Ono (left) and Kashimadai (right) plotted against the squared height of the tidal envelope p^2 at Miyato

wave set-up w, as indicated by that the residual varies in co-phase with the tidal envelope at Miyato. Even though the tide largely dissipates by Kashimadai, and its visual amplitude is about 1/4 of that at Ono, the tidal wave set-up is approximately the same at both locations.

But how exactly does the wave set-up upriver depend on tidal forcing at the mouth? In Fig. 3.6, lower panes, the residuals $\hat{r}(t)$ at Ono (lower left) and Kashimadai (lower right) are mapped against a squared height of the tidal envelope at Miyato p^2. Apparently, $\hat{r}(t)$ has a component proportional to p^2. Therefore, we can assume that

$$w(x, p) = \beta_2(x)p^2, \tag{3.4}$$

and sub-tidal MRS equation (3.2) becomes

$$r(x, t) = \beta_0(x) + \beta_1(x)s(t) + \beta_2(x)p^2(t) + \epsilon, \tag{3.5}$$

with ϵ being the fit residual. Coefficients $\beta_{0,1,2}$ are now found by fitting a ST-MRS r recorded at a particular in-river location with a constant function, the ST-MSL s, and the squared height of the tidal envelope p^2 measured near the river mouth. It is expected that $\beta_1(x)$ diminishes upriver starting from $\beta_1(0) = 1$ at the mouth, while the wave set-up profile $\beta_2(x)$ starts with 0 at the mouth, reaches its peak at some upriver distance, and then diminishes to zero. Coefficient β_0 will not be interpreted, since it takes on non-zero averages in both the measurements and the fitting functions.

Applying decomposition (3.5) to the records in Yoshida yields $\beta_1 = 1.03$ and $\beta_2 = 0.00$ at Nobiru, $\beta_1 = 0.95$ and $\beta_2 = 0.061$ at Ono,[3] and $\beta_1 = 0.54$ and $\beta_2 = 0.046$ at Kashimadai (Table 3.1). As the tidal range p within the record time varied from 0.55 to 1.67 m, our estimated tidal set-up $w = \beta_2 \cdot p^2$ at Ono varied from 0.018 (0.06×0.55^2) to 0.17 m; and that at Kashimadai varied from 0.014 to 0.13 m.

The 2011 Tohoku tsunami entered the Yoshida River with a 6-m high first wave, followed by a train of approximately 4-m high (peak-to-trough) waves (Fig. 1.2). Extrapolating the above $w(p)$ dependance to the 2011 tsunami event, a train of waves

Table 3.1 Gauging stations in Yoshida; their distances from the mouth; fraction β_1 of ST-MSL variations carried into ST-MRS; wave set-up coefficient β_2; and estimated minimal and maximal tidal set-up heights in the records, at each station

Station	rkm	β_1	β_2, m^{-1}	w_{min}, cm	w_{max}, cm
Nobiru	0.5	1.03	0.0	0	0
Ono	4.2	0.95	0.061	1.8	17
Kashimadai	9.0	0.54	0.046	1.4	13

---

[3]Fitting a record with 1 and s alone and with 1, s, and p^2 jointly can result in different coefficients $\beta_{0,1}$, because in mathematical terms, 1, s, and p^2 are not mutually orthogonal.

with a 4-m peak-to-trough range should elevate MRS through $0.06 \times 4^2 \approx 1$ m at Ono, and $0.046 \times 4^2 \approx 0.74$ m at Kashimadai. It will be demonstrated in Sect. 4.5, however, that a wave train in a tsunami frequency band causes about twice as high a wave set-up as the tide does. Our tsunami set-up estimates are thus 2 m at Ono, and 1.5 m at Kashimadai, which agrees with observations (Fig. 1.2).

3.2.2   Tidal Wave Set-Up in the Columbia River, USA

The Columbia River on the west coast of North America provides a contrast to the smaller, steeper, and more confined rivers in Japan. The Columbia River has a 3.5 km wide mouth bordered with jetties; an approximately 80 rkm long estuary which widens to 15 km in its lower reach; extensive tidal flats; a tiny bed slope of 3.5 cm per km for the lower 200+ rkm; and a 170 rkm long, 10–15 m deep dredged channel from the mouth to Portland (Yeh et al. 2012; Jay et al. 2015). Tides penetrate into the river until stopped by the Bonneville Dam at 234 rkm. Yeh et al. (2012) discuss simulations of a hypothetical Cascadia mega-tsunami invading the Columbia River. The tsunami 5.5 m high at the mouth becomes only 0.8 m high by the beginning of the fluvial reach at Skamokawa. The tsunami waveform also changes dramatically. Not only are the higher frequency components filtered out, but at all virtual gages past Skamokawa, the river surface remains elevated upon the tsunami's arrival and for the entire simulation time of 15 h after the wave has entered the river, which suggests the presence of backwater in the time histories.

In spite of a very different river morphology, the tidal set-up in Columbia is no less apparent than in Yoshida. Below, in the same manner as above, we examine tidal records from March through October, 2013 at seven NOS (National Ocean Service, NOAA) gages between the river mouth and Vancouver/Portland (Fig. 1.5). Figure 3.7 shows tidal records and corresponding ST-MRS at all stations. Visually, very little correlation between ST-MRS and tidal envelope is present at Hammond, some traces of such correlation appear in Astoria, while at the more upstream stations, ST-MRSs tend to go up whenever the tidal range increases.

To quantify accumulation of the tidal set-up, we apply decomposition (3.5) to each record upstream of Hammond. The most downstream record at Hammond provides a proxy for ocean forcing. ST-MRS at Hammond $r_0(t)$ is treated as ST-MSL s, and ocean forcing is evaluated by tidal envelope height at Hammond $p(t)$.

Figure 3.8 shows the tidal range p at Hammond, and ST-MRS residuals $\hat{r}(x_j, t) = r(x_j, t) - \beta_1(x_j) \cdot r_0(t)$, $j = 1 - 6$ at the other six stations. The residuals vary in co-phase with the tidal envelope. Proceeding as with Yoshida records, the residual ST-MRS variations $\hat{r}(x_j, t)$ are mapped against squared tidal range p^2 at Hammond, in Fig. 3.9. The residuals show the presence of the wave set-up component proportional to p^2, and the contribution from the upstream river conditions. The latter contribution is small in the lower estuary, but greatly increases upriver past Skamokawa. Once the wave set-up coefficients $\beta_2(x_j)$ are determined, the backwater height at each station can be evaluated with Eq. (3.4). The results

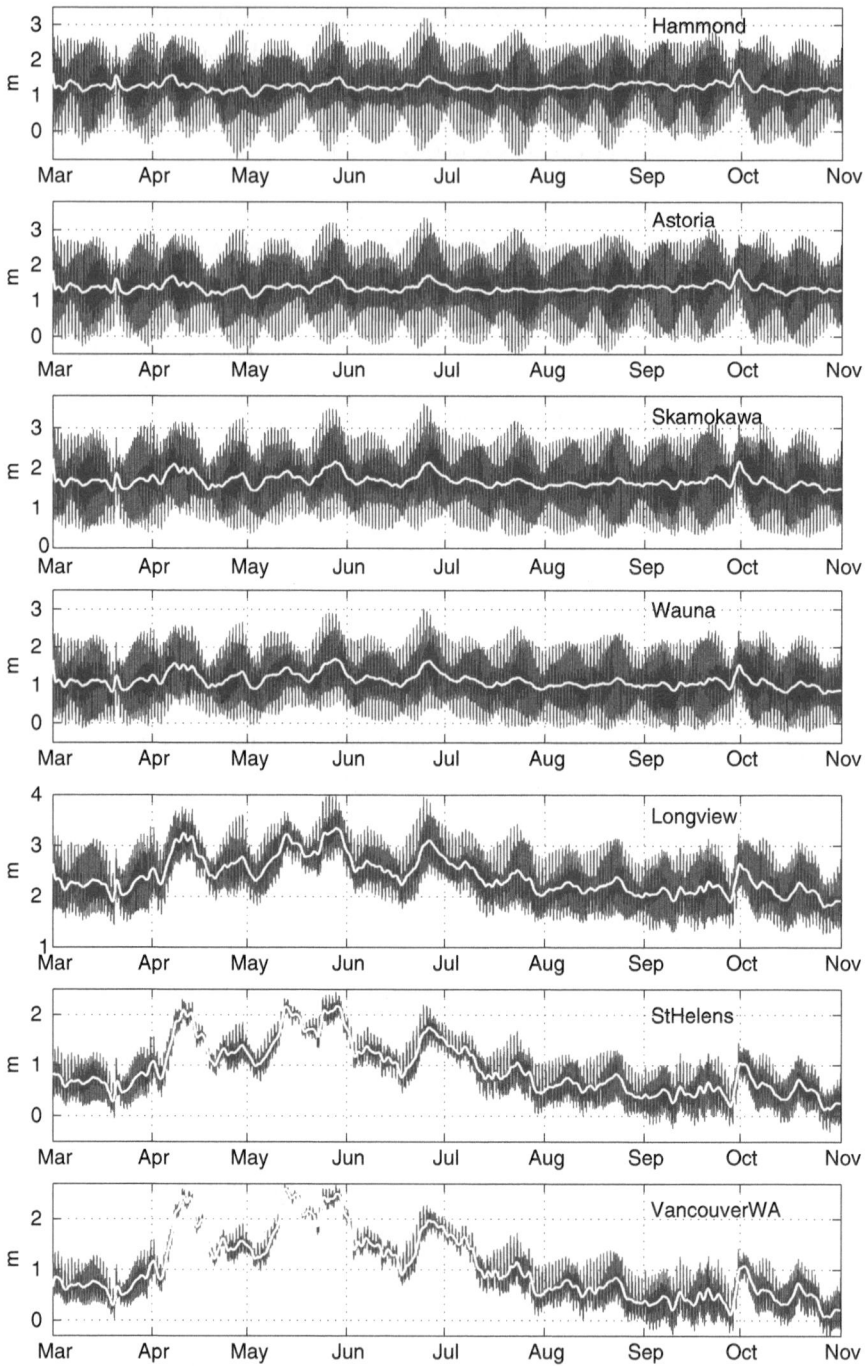

Fig. 3.7 Water level measurements along the Columbia river from March through October, 2013 at seven NOS gauging stations (gray), and ST-MRS in each record (white)

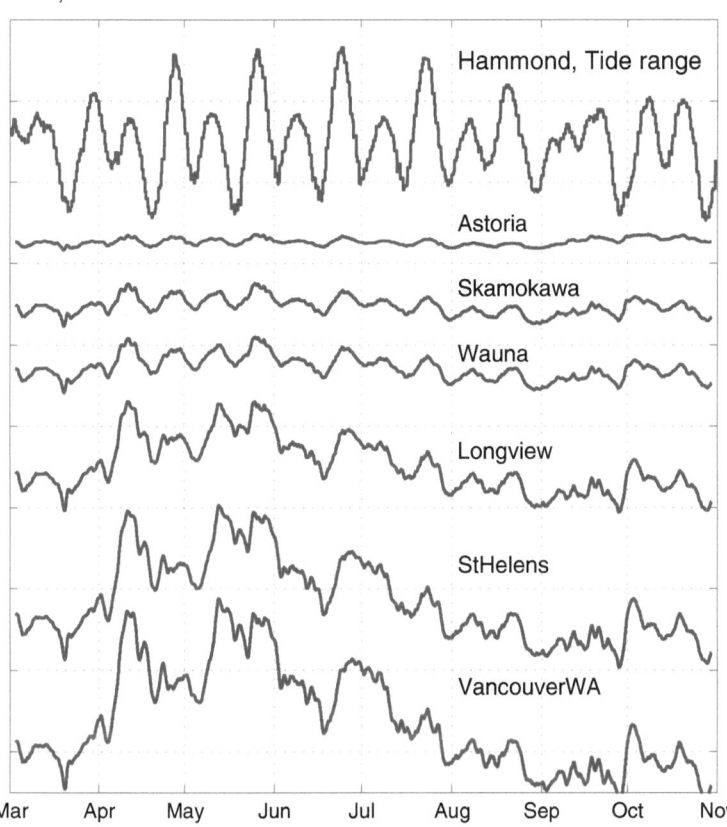

Fig. 3.8 Tidal range at Hammond; ST-MRS variations with response to ST-MSL variations removed at six upriver stations, from 2013/03/01 to 2013/10/31. Vertical scale is 1 m between grid lines. No common vertical reference among different plots

are summarized in Table 3.2, which lists coefficients β_1 and β_2 at each station, the estimates of the maximal backwater height variation during the record time

$$\Delta w = \max w - \min w = \beta_2 \cdot \left(\max p^2 - \min p^2 \right),\qquad(3.6)$$

and the r.m.s. deviation of the backwater height

$$\sigma w = \sqrt{< (w - < w >)^2 >}.\qquad(3.7)$$

The river stage excursion in response to varying strength of the ocean forcing grows away from the ocean, within the first 170 rkm of the Columbia river. At more than 100 rkm from the ocean, the river stage moves up by as far as 0.6 m when the ocean tides change from neap to spring. Both the tide and the tidal wave set-up are stopped by the Bonneville Dam at 234 rkm.

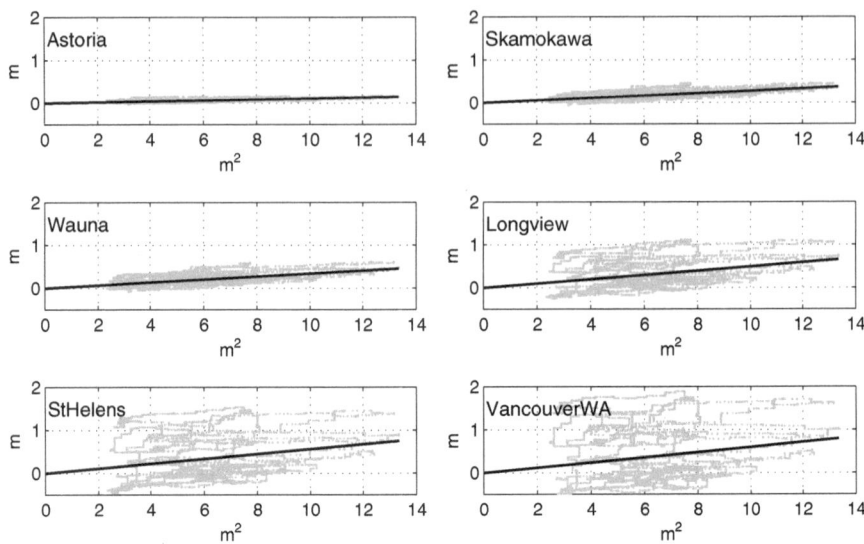

Fig. 3.9 Residual ST-MRS variations $\hat{r}(x_j, t)$ vs. the squared tidal range p^2 at Hammond (gray dots), and fit with a line $\beta_2 \cdot p^2$ (black line)

Table 3.2 Observation stations; their distances from the mouth; fraction β_1 of ST-MSL variations carried into ST-MRS; wave set-up coefficient β_2; estimated range of the backwater height variation (3.6); and r.m.s. variation of the backwater height (3.7), at each station

	Hammond	Astoria	Skamokawa	Wauna	Longview	St. Helens	Vancouver, WA
rkm	15	29	54	67	107	139	170
β_1	1.0	0.96	0.96	0.94	0.87	0.77	0.70
β_2, m^{-1}	0.0	0.011	0.027	0.034	0.049	0.056	0.059
Δw, cm	0	11.6	29.9	37.2	53.7	61.2	64.9
σ_w, cm	0	2.5	6.4	7.95	11.5	13.1	13.9

3.2.3   Don't Tidal Scientists Know All That?

Indeed, the Lower Columbia River and Estuary have long been a study area for tidal scientists. Tidal analysis operates in terms of harmonic or wavelet analysis, when a water level record is decomposed into predefined functions, usually sines and cosines at prescribed frequencies—harmonic constituents (Pugh and Woodworth 2014). Each such constituent is considered a physical entity with its own name, cause, and behavioral characteristics. In a complex (non-linear) medium such as a river, the constituents interact with each other and give birth to more constituents. A physical phenomenon is then interpreted through behavior of the constituents presumably comprising this phenomenon. Therefore, the phenomenon under study must first be identified by its period/frequency, and matched with the corresponding constituent(s).

Tidal wave set-up manifests itself by a river's stage variation over a synodic cycle. Thereby from the tidal analysis perspective, the tidal set-up is a process with a 15-day period represented by a harmonic constituent named MSf. An MSf cause as given by Jay et al. (2015) presents a conventional interpretation of the tidal set-up: "MSf represents a tide-flow interaction—there is greater tidal friction on the flow during spring tides. In addition, there is an additional discharge on spring tides that compensates for the larger landward Stokes drift on springs. For both of these reasons, river slopes (therefore, upriver water levels) must be higher to discharge the same amount of water from land." Jay et al. (2015) have performed tidal analysis of water level records along the Columbia river, in particular, to estimate distribution and variation of tidal wave set-up. The phenomenon has been identified by spectral peaks at a 15-day period, which are absent in the ocean, emerge at Astoria, get stronger at Skamokawa, and become "very prominent" at St. Helens. In Fig. 3.10, we reproduce the wave set-up variation along the river given by an amplitude of MSf constituent in tidal records, as in Fig. 7 in Jay et al. (2015). We also show the r.m.s. variation σ_w of the backwater height, and a half of its variation range $0.5\Delta w$, as computed here. We observe a good agreement between the backwater profile shapes. A sine function with a 16-cm amplitude (the peak MSf amplitude) has a $16/\sqrt{2} = 11.3$ cm r.m.s. value, which is close to our values for σ_w. However, anywhere in the river, the backwater height range computed here is twice as large as the height of the MSf constituent found by Jay et al. The reason for this discrepancy becomes obvious once we look at the tidal envelope at Hammond, in Fig. 3.8, top. The synodic variation of the tidal envelope there does not follow the sine law. Moreover, this envelope does not even form a periodic function. Consequently, the river's response does not follow the sine law either, and cannot be reduced to the MSf constituent.

For this reason, we attempt somewhat different approaches to investigating long-wave dynamics in rivers. We don't expect anything harmonic in a river, and adhered to this condition with our analytical solution for the backwater profile in Sect. 2.3.

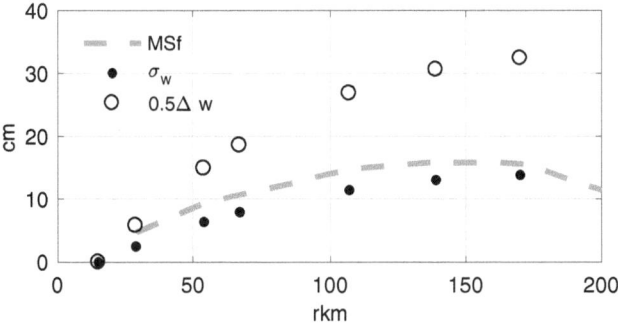

Fig. 3.10 MSf amplitude along the Columbia river as computed by Jay et al. (2015); r.m.s. variation of the backwater height σ_w, and a half of the range of backwater height variation $0.5\Delta w$, as found here

We expect the non-linear mechanisms to affect the entire wave form directly, rather than to operate on its parts. How then can the physical processes under study be separated from other processes? There is no simple answer to this question anymore. In the present case, it was hypothesized that the tidal set-up could be extracted from the measurements as a fraction of ST-MRS proportional to the tide intensity at the mouth. More in-depth analysis of the wave set-up phenomenon will relay on "observations" supplied by numerical simulations.

References

Atwater, B. F., Musumi-Rokkaku, S., Satake, K., Tsuji, Y., Ueda, K., & Yamaguchi, D. K. (2015). *The orphan tsunami of 1700 – Japanese clues to a parent earthquake in North America* (2nd ed., p. 135). Seattle: University of Washington Press. U.S. Geological Survey Professional Paper 1707. https://doi.org/10.3133/pp1707.

Buschman, F. A., Hoitink, A. J. F., van der Vegt, M., & Hoekstra, P. (2010). Subtidal flow division at a shallow tidal junction. *Water Resources Research, 46*, W12515. https://doi.org/10.1029/2010WR009266.

Hoitink, A. J. F., & Jay, D. A. (2016). Tidal river dynamics: Implications for deltas. *Reviews of Geophysics, 54*, 240–272. https://doi.org/10.1002/2015RG000507.

Jay, D. A., & Flinchem, E. P. (1997). Interaction of fluctuating river flow with a barotropic tide: A demonstration of wavelet tidal analysis methods. *Journal of Geophysical Research, 102*(C3), 5705–5720.

Jay, D. A., Leffler, K., Diefenderfer, H. L., & Borde A. B. (2015). Tidal-Fluvial and Estuarine processes in the lower Columbia River: I. Along-channel water level variations, Pacific Ocean to Bonneville Dam. *Estuaries and Coasts, 38*, 415–433. https://doi.org/10.1007/s12237-014-9819-0.

LeBlond, P. H. (1979). Forced fortnightly tides in shallow waters. *Atmosphere-Ocean, 17*(3), 253–264.

Pugh, D., & Woodworth, P. (2014). *Sea-level science: Understanding tides, surges, tsunamis and mean sea-level changes* (2nd ed., p. 395). Cambridge: Cambridge University Press.

Sassi, M. G., & Hoitink, A. J. F. (2013). River flow controls on tides and tide-mean water level profiles in a tidal freshwater river. *Journal of Geophysical Research, Oceans, 118*, 4139–4151. https://doi.org/10.1002/jgrc.20297.

Tanaka, H., Ishino, K., Nawarathna, B., Nakagawa, H., Yano, S., Yasuda, H., et al. (2008). Field investigation of disaster in Sri Lankan rivers caused by the 2004 Indian Ocean Tsunami. *Journal Hydroscience and Hydraulic Engineering, 26*(1), 91–112.

Yeh, H., Tolkova, E., Jay, D., Talke, S., & Fritz, H. (2012). Tsunami Hydrodynamics in the Columbia River. *Journal of Disaster Research, 7*(5), 604–608.

Chapter 4
Tsunami and Tidal Set-Up in Rivers: A Numerical Study

Highlights *Quantifying rivers and their responses. Simulation arrangements. A tsunami and a tide in a numerical river: alike and different. Patterns, relations, numbers. Effects of the bed shape, wave period, riverine current, wave amplitude. Two metrics for the wave decay with distance travelled. Surprises of high water marks: a larger wave dissipates slower. A distance to the wave set-up peak as a scale for wave penetration in a river. Can tidal observations be used for predicting a tsunami's behavior?*

The questions asked through our investigation are: what defines long wave behavior in a river? How much does the wave attenuate with distance, or in other words, how far upriver will a tsunami penetrate? How will the tsunami intrusion appear to an observer located at 1 rkm? At 10 rkm? At 25 rkm? A river's primarily parameters—depth, breadth, bottom slope, bottom roughness, freshwater current—describe what the river is made of. These parameters do not answer our questions directly. In some yet unknown way, the primarily parameters define the river's "behavioral" parameters which characterise the wave effects.[1] We will find that raising the river's mean stage during the wave intrusion affects the entire spectrum of wave effects, and that the parameters characterizing the wave set-up also describe the wave attenuation with an upriver distance.

---

[1]For an analogy, primary parameters specifying a particular frictionless system comprised of a rigid body suspended from a coil spring include the shape and mass of the body, the density and Young's modulus of the spring's material, spring length, and number and radius of the spring's coils. These primary parameters yield one "behavioral" parameter—a period of oscillation—characterising the system's motion along the vertical axis.

© The Author(s) 2018
E. Tolkova, *Tsunami Propagation in Tidal Rivers*, SpringerBriefs
in Earth Sciences, https://doi.org/10.1007/978-3-319-73287-9_4

4.1   River Models

The quest for a river's "behavioral" parameters will rely on numerical simulations of a very simple scenario: a harmonic wave intruding in a 1-D river which has a uniform bed slope $\beta_0 = -d_x$, a uniform bottom roughness n, a uniform depth h_0, and a uniform cross-sectional area (in an undisturbed state). The latter amounts to a uniform freshwater current u_0 related to the other parameters by Eq. (2.28): $\beta_0 = n^2 u_0^2 / h_0^{4/3}$. The simulations will use a fixed river depth $h_0 = 5\,\text{m}$, in understanding that the SWE solutions can be scaled with respect to depth as discussed in Sect. 2.1.

We are interested in finding an established solution of the SWE in our simplified river

$$h_t + (uh)_x = 0, \quad u_t + uu_x + gh_x = -\beta_0 + f, \tag{4.1}$$

given harmonic surface motion at the mouth

$$\eta(t, 0) = a \sin(2\pi t / T), \tag{4.2}$$

and a uniform flow state at the upstream end

$$\eta(t, x_{end}) = \beta_0 \cdot x_{end}, \quad u(t, x_{end}) = u_0. \tag{4.3}$$

The established solution is expected to be periodic (but not harmonic) with a period of an external forcing.

Our modeling study of the river responses to the periodic ocean forcing will include nine river models with three different values of bed roughness n and three different bed slopes β_0. Each river model will be penetrated by six waves characterized by three values of forcing amplitude a, and two forcing periods representing tsunami and tidal bands: $T_1 = 1\,\text{h}$, and $T_2 = 12.4\,\text{h}$ (M2 tidal constituent).

4.2   Quantifying a River's Response

A river's response to the periodic ocean forcing comprises accumulation of the backwater and concurrent dissipation of the intruding wave. Herein, bars denote averaging over the wave period T. The wave-induced backwater, which is the water accumulated between an undisturbed river's stage and its mean stage under propagating wave (Fig. 3.4), will be characterized by

1. the height of the backwater at a distance x from the mouth:

$$\xi_S(x) = \overline{\eta}(x) - \eta_0(x); \tag{4.4}$$

2. the backwater accumulation distance L_ξ, given by a distance from the mouth to the point of the maximal wave set-up;

3. the maximal backwater height:

$$H_\xi = \xi_S(L_\xi) = \max(\xi_S(x)); \qquad (4.5)$$

4. the volume of the backwater (per unit river width) between the mouth and an upriver position x:

$$V(x) = \int_0^x \xi_S(\tilde{x})d\tilde{x}; \qquad (4.6)$$

5. $V_\infty = V(x_{end})$—the total volume of water accumulated above the original river stage;
6. $V_{L_\xi} = V(L_\xi)$—the volume of water accumulated between the mouth and the point of the maximal wave set-up.

The wave action at an upriver position x will be characterized by

1. the variance of the observed surface motion in a sequence of rise-and-fall about its mean elevation:

$$w(x) = \overline{(\eta(x) - \overline{\eta}(x))^2} = \overline{\xi_W^2}. \qquad (4.7)$$

At the mouth, where the surface motion follows a sine with an amplitude a, $w(0) = a^2/2$.
2. the maximal elevation of the river surface above its original level:

$$\xi_{max}(x) = \max(\eta(x, t)) - \eta_0(x). \qquad (4.8)$$

This is a value inferred from high water marks in post-tsunami field surveys. It might be mistaken for the wave amplitude, should somebody assume that the river surface oscillates about its original level. At the mouth, though, it is the wave amplitude: $\xi_{max}(0) = a$.

4.3 Simulation Arrangements and Results

The simulations are performed with an open-source model *Cliffs* adapted, in particular, for modeling river flows. The model solves the fully-nonlinear SWE with a choice of 1-D or 2-D configuration, Cartesian or spherical coordinates, initial conditions and/or boundary forcing, constant or varying grid spacing (Tolkova 2014; Lynett et al. 2017). The bed friction is in Manning formulation.

Three bottom slopes β_0 (0.11, 0.19, and 0.29 m/km), paired with three Manning coefficients n (0.03, 0.04, and 0.06) each, yield nine river models entertained in the following study, with a respective riverine current $u_0 < 0$ computed according to (2.28), $h_0 = 5$ m. Each river domain is 410 km long and contains 2367 computational nodes. The first 1667 nodes cover the first 50 rkm with a 30 m spacing. Then the grid spacing gradually increases and reaches 999 m between the last two nodes, where the bed elevations for the three slope are 38 m, 72 m, and 115 m, respectively, above the surface level at the mouth. The domain was initialized with a uniform flow state at $t = 0$, that is, the flow velocity at all nodes was set to u_0, and the surface elevation was set to $\eta_0(x) = \beta_0 \cdot x$. No wave effects penetrate to the upriver end of the domain, where the boundary conditions are prescribed by the uniform flow state (4.3). At the mouth, the surface elevation used to represent an incoming wave[2] varied as $2a \sin (2\pi t/T)$ for $t > 0$, with three amplitudes a and two periods T (1 and 12.4 h) used in each river. The waves at these two periods will be referred to as the shorter and the longer waves, or as "tsunami" and "tide". Should the domain be a frictionless deep flat channel, the latter forcing would induce a simple wave of amplitude a and flow velocities varying in co-phase with the surface elevation. In our rivers, however, the velocity at the mouth would follow an unknown (until computed) law, and the amplitude of the resulting wave will deviate from a, being slightly different in different rivers. In each case, an exact value of this amplitude is obtained from the simulations. Flow reversal at the river mouth took place in fifteen "tsunami" scenarios (marked with the star in Table 4.1), and did not happen in any "tidal" scenario. Simulated surface elevation and current in each scenario are saved with an interval 10 s (for $T = 1$ h) or 60 s (for $T = 12.4$ h) at every 3-d node starting from the mouth, for the first 100 rkm.

Thereby, this numerical experiment comprises 54 scenarios of 6 harmonic waves, characterized by 3 different amplitudes and 2 periods, intruding in 9 rivers characterized by 3 slopes and 3 friction coefficients. The river parameters in each scenario are listed in the left three columns in Table 4.1; wave amplitudes at the mouth and the backwater accumulation parameters (4.4)–(4.6) computed in each simulation are listed in the next four columns for a wave in the tsunami frequency range ($T = 1$ h), and in the right four columns for a wave in the tidal range ($T = 12.4$ h).

Which wave behaviors emerge from these numbers?

4.4   First Look at the Solutions

Two numerical solutions in a river model with $h_0 = 5$ m, $\beta_0 = 0.11$ m/km, $n = 0.04$ s $\cdot$ m$^{-1/3}$, and $u_0 = -0.75$ m/s (6-th line in Table 4.1) are displayed in Figs. 4.1 and 4.2. The pictured solutions represent waves 2.52 m high at the mouth with periods 1 h and 12.4 h, respectively. The top two panels show along-river surface profiles in an established solution at the moments when a wave crest or wave trough, respectively, passes through the mouth. Drawn atop the profiles are the river's mean stage, and the initial stage. Lower rows of plots show time histories of the surface elevation and flow velocities from the start of the simulation,

---

[2]That is, to compute an incoming Riemann invariant.

Table 4.1 Input and output: river, wave, and backwater accumulation parameters

River			$T = 1\,\mathrm{h}$				$T = 12.4\,\mathrm{h}$			
β_0		$-u_0$	a	L_ξ	H_ξ	V_∞	a	L_ξ	H_ξ	V_∞
(m/km)	n	(m/s)	(m)	(km)	(m)	$(10^3\,\mathrm{m}^2)$	(m)	(km)	(m)	$(10^3\,\mathrm{m}^2)$
0.11	0.03	1	0.56	5.8	0.09	1.8	0.56	7.9	0.03	0.7
	0.03	1	1.14*	6.2	0.33	7.5	1.2	7.9	0.13	3.3
	0.03	1	2.41*	7.5	1	29.5	2.44	7.9	0.5	14.5
	0.04	0.75	0.59	5.3	0.11	2.2	0.58	7.7	0.03	0.8
	0.04	0.75	1.19*	5.9	0.4	9	1.24	7.7	0.14	3.7
	0.04	0.75	2.52*	7.1	1.13	32.7	2.53	7.9	0.57	16.4
	0.06	0.5	0.63*	4.7	0.15	2.8	0.6	7.4	0.04	0.9
	0.06	0.5	1.28*	5.6	0.5	10.8	1.29	7.4	0.17	4.2
	0.06	0.5	2.7*	6.3	1.25	33.8	2.63	8	0.68	19.4
0.19	0.03	4/3	0.59	3.7	0.07	0.9	0.56	4.8	0.02	0.3
	0.03	4/3	1.18	3.8	0.26	3.5	1.2	4.7	0.1	1.6
	0.03	4/3	2.41*	4.3	0.86	14.5	2.45	4.5	0.41	6.8
	0.04	1	0.62	3.4	0.08	1	0.58	4.8	0.03	0.4
	0.04	1	1.23	3.7	0.31	4.2	1.25	4.7	0.11	1.8
	0.04	1	2.52*	4.3	1	17	2.55	4.6	0.45	7.6
	0.06	2/3	0.66	3.2	0.11	1.3	0.61	4.7	0.03	0.4
	0.06	2/3	1.31*	3.5	0.41	5.4	1.3	4.6	0.13	2
	0.06	2/3	2.7*	4.1	1.16	19.5	2.65	4.7	0.52	8.8
0.29	0.03	5/3	0.6	2.5	0.06	0.5	0.55	3.2	0.02	0.2
	0.03	5/3	1.21	2.5	0.22	1.9	1.18	3.1	0.09	0.9
	0.03	5/3	2.43*	2.7	0.75	8	2.42	2.9	0.36	3.9
	0.04	1.25	0.64	2.4	0.07	0.6	0.58	3.2	0.02	0.2
	0.04	1.25	1.27	2.5	0.25	2.3	1.24	3.2	0.1	1
	0.04	1.25	2.52*	2.8	0.87	9.5	2.53	3	0.39	4.3
	0.06	5/6	0.69	2.3	0.09	0.7	0.61	3.2	0.03	0.2
	0.06	5/6	1.36*	2.4	0.33	2.9	1.3	3.2	0.11	1.1
	0.06	5/6	2.69*	2.9	1.05	11.5	2.65	3.1	0.44	4.9

Stars next to incoming wave amplitude mark wave scenarios with flow reversals

as well as the respective initial surface elevations and currents. The time histories are displayed at the mouth, at 7.5 rkm, and at 15 rkm. Common to both solutions, at some upriver distance, the river surface remains elevated above the initial level at all times, regardless of the wave phase (peak or trough) at the mouth. Different between the shorter ($T = 1\,\mathrm{h}$) and longer ($T = 12.4\,\mathrm{h}$) waves,

- the shorter wave (tsunami) elevates the river's mean stage to a much greater height than the longer wave (tide). Even with a tsunami trough at the mouth, the river stage does not recede below its initial level upstream a 2 rkm mark (Fig. 4.1, second panel from the top); whereas at the low tide, the river stage drops below the initial level for the first 10 rkm (Fig. 4.2, second panel from the top).

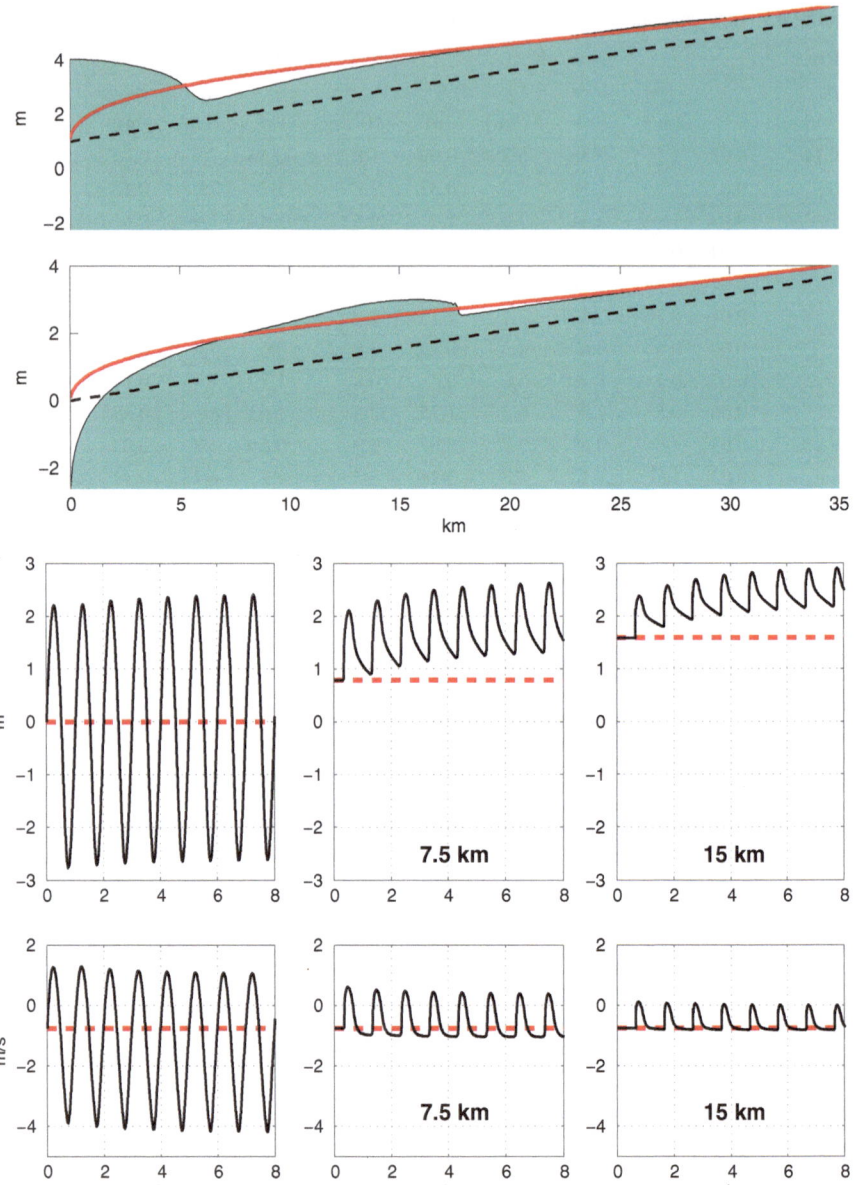

Fig. 4.1 Top two panels: a river's view in the $x - z$ plane when the wave crest (upper panel) or trough (lower panel) passes by the mouth. The river is invaded by a harmonic wave with an amplitude 2.52 m and a period $T = 1$ h; solid red—a river's mean stage set by the wave; dashed black—an undisturbed river stage. Lower two rows: solid black—time histories of the surface elevation (upper row) and flow velocity (bottom row) at the mouth, at 7.5 rkm, and at 15 rkm; dashed red—an undisturbed surface level or flow speed, respectively. River's parameters: $h_0 = 5$ m, $\beta_0 = 0.11$ m/km, $n = 0.04$ s $\cdot$ m$^{-1/3}$, and $u_0 = -0.75$ m/s

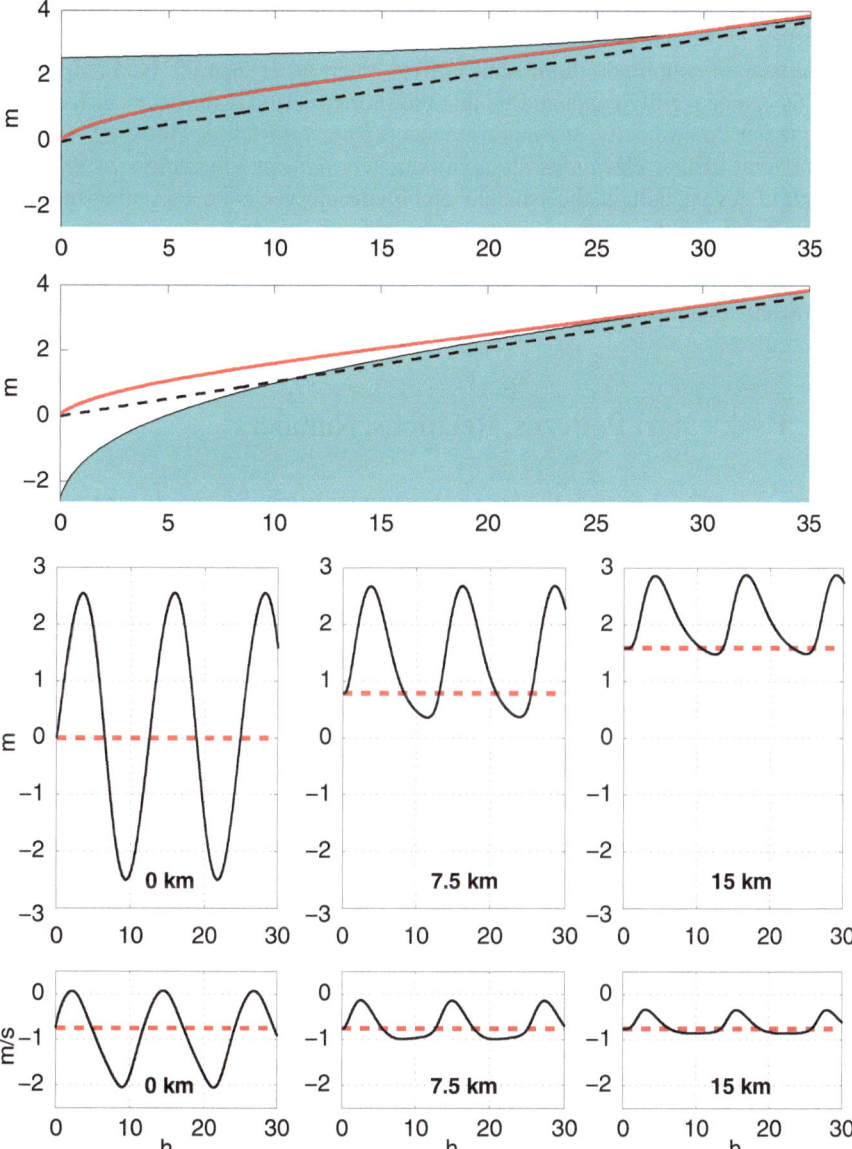

Fig. 4.2 Same as in Fig. 4.1, for a wave period $T = 12.4\,\mathrm{h}$

- it takes four tsunami periods, but less than one tidal period, to fill up the river and establish the elevated river stage.
- with the same wave height at the mouth, tsunami-induced currents are much stronger than tidal currents (compare lower left plots in Figs. 4.1 and 4.2). Tidal currents noticeably deviate from a harmonic law even at the mouth, where the

surface motion is harmonic. The peak tidal inflow current precedes the peak surface elevation, while these of the tsunami occur practically simultaneously.

- the tsunami amplitude diminishes upriver much faster than the tidal amplitude. This can be readily explained by that the shorter-period wave enters a river with a greater flow velocity, and therefore meets a greater friction. However,
- *maximal surface elevations are approximately the same regardless of the wave period* (event though the tsunami amplitude upriver is smaller than the tidal amplitude, but the wave set-up under the tsunami is higher).

Next, we'll attempt to quantify these and other behaviors in terms of our Output Parameters obtained in a set of simulations of different waves in different rivers.

4.5   Backwater: Patterns, Relations, Numbers

Figures 4.3 and 4.4 display the backwater accumulation distance L_ξ, the maximal height H_ξ, and the total backwater volume V_∞ as functions of the wave amplitude at the mouth, obtained in 27 scenarios with $T = 1$ h (Fig. 4.3) and in 27 more—with $T = 12.4$ h (Fig. 4.4). Data markers are color-coded according to bed slopes,[3] and shaped according to bed friction.

Clear patterns emerge in these data:

(a) In the same river and for the same wave amplitude at the mouth, the height and the total volume of the backwater is about twice as high in respond to a wave train at a tsunami period (Fig. 4.3) than that at a tidal period (Fig. 4.4).

(b) In the tsunami frequency range, the backwater accumulation distance depends primarily on the bed/surface slope, and only mildly—on the bed roughness and the wave amplitude (Fig. 4.3a). In the tidal range, this distance depends entirely on the bed/surface slope, being roughly independent of other river or wave conditions considered (Fig. 4.4a).

(c) The milder the bed slope, the greater both the backwater accumulation distance and the total backwater volume.

(d) The greater the wave amplitude and the milder the bed slope, the greater the backwater height, with little to no dependance on the bed roughness.

(e) Without flow reversal ($T = 12.4$ h), the backwater accumulation distance is inversely proportional to the bed slope, with little dependance on other parameters (see Fig. 4.5 for a nearly constant product $L_\xi \cdot \beta$ in all 27 "tidal" scenarios.)

---

[3]In rivers with uniform depth considered here, we cannot distinguish between a bed slope and a surface slope. In a more general case, a surface slope might constitute a more meaningful parameter than a bed slope.

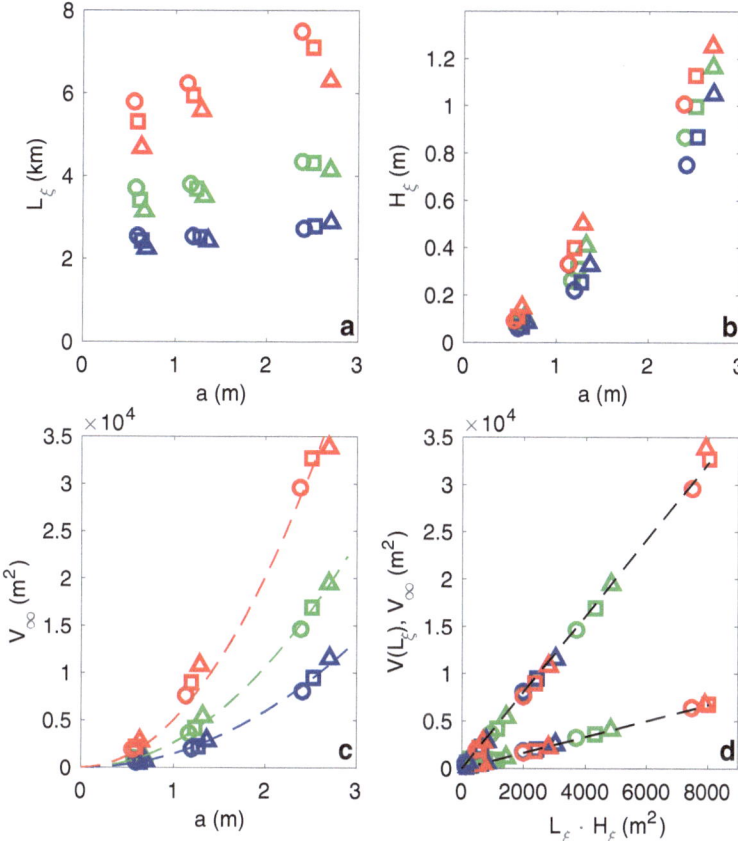

Fig. 4.3 (a) The distance L_ξ from the river mouth to the point of the maximal wave set-up/backwater accumulation, as a function of a wave amplitude a at the mouth; (b) the maximal backwater height H_ξ as a function of the wave amplitude a; (c) the total backwater volume V_∞ per unit river width, as a function of the amplitude a; (d) the total backwater volume in the river (per unit width) V_∞ and that between the mouth and the point of the maximal wave set-up $V(L_\xi)$, vs. $L_\xi \cdot H_\xi$. Colors refer to different bed slopes: 0.11 m/km (red), 0.19 m/km (green), and 0.29 m/km (blue); markers—to different Manning coefficients: $n = 0.03$ (circles), $n = 0.04$ (squares), and $n = 0.06$ (triangles). Dashed lines: least-square fit with parabolas $y = kx^2$ and straight lines $y = kx$. Wave period $T = 1$ h

(f) At both periods considered, the total volume of the accumulated water is approximately proportional to the square of the wave amplitude at the mouth, with a factor depending on the bed slope and the wave period, but independent of the bed roughness or a riverine current: $V_\infty = k(\beta, T) \cdot a^2$.

(g) In all scenarios, the total backwater volume, as well as the backwater volume between the mouth and the point of the maximal wave set-up, is directly

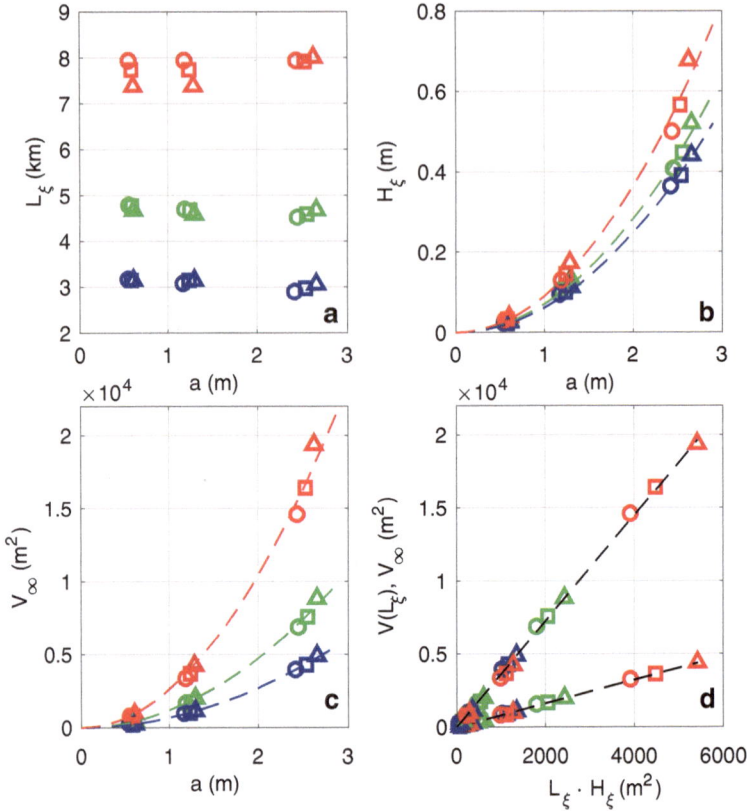

Fig. 4.4 The same as in Fig. 4.3, for wave period $T = 12.4$ h

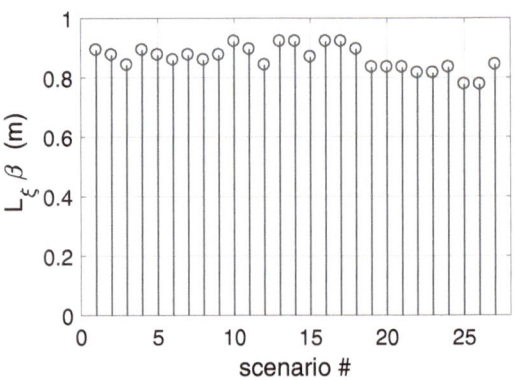

Fig. 4.5 The products of the backwater accumulation distance (in km) and the bottom slope (in m/km) $L_\xi \cdot \beta$ in 27 "tidal" scenarios

proportional to the product of the backwater accumulation distance and the maximal wave set-up height:

$$V_\infty = k_1 \cdot H_\xi \cdot L_\xi, \quad V_{L_\xi} = k_2 \cdot H_\xi \cdot L_\xi \qquad (4.9)$$

with $k_1 = 4.0$ and $k_2 = 0.84$ if estimated with $T = 1\,\mathrm{h}$ simulation dataset, and $k_1 = 3.6$ and $k_2 = 0.8$ with $T = 12.4\,\mathrm{h}$ dataset. Consequently, a fraction of all backwater located downstream of the point L_ξ is nearly constant in all scenarios:

$$\frac{V_{L_\xi}}{V_\infty} = \frac{k_2}{k_1} \approx 0.22 \qquad (4.10)$$

By correlation with Eq. (2.52), the last pattern yields an important conclusion: the unknown quantity α introduced in the course of analytical manipulations (2.49)–(2.52) is, in fact, a constant.

4.5.1 Backwater Curve

Given the definition of the backwater volume (4.6), the data pattern G also suggests, that backwater curves $\xi_S(x)$ in all considered rivers have the same shape stretched horizontally according to a river's L_ξ and vertically according to its H_ξ:

$$\xi_S(x) = H_\xi \cdot \phi(x/L_\xi), \quad \phi(1) = 1. \qquad (4.11)$$

To check on this hypothesis, all computed backwater curves normalized by their respective heights are plotted as a function of an upriver distance normalized by a respective accumulation distance. The wave set-up starts from a non-zero (though small) value at the mouth—a consequence of our lack of a transition from a river stream to a larger reservoir (making such a transition would add complexity and cost additional parameters to characterize the system). All 27 such curves corresponding to $T = 1\,\mathrm{h}$ are plotted in the same axes in Fig. 4.6, upper pane, while 27 curves for $T = 12.4\,\mathrm{h}$ are plotted in the lower pane. Each set of curves is overlaid with an analytical curve calculated with expressions (2.49)–(2.51), with $\alpha = 0.22$. A constant α was estimated by solving $\kappa(\alpha) = k_1 = 3.6$, where $\kappa(\alpha)$ is defined by (2.52). All simulated backwater profiles closely agree with each other and with the analytical estimate.

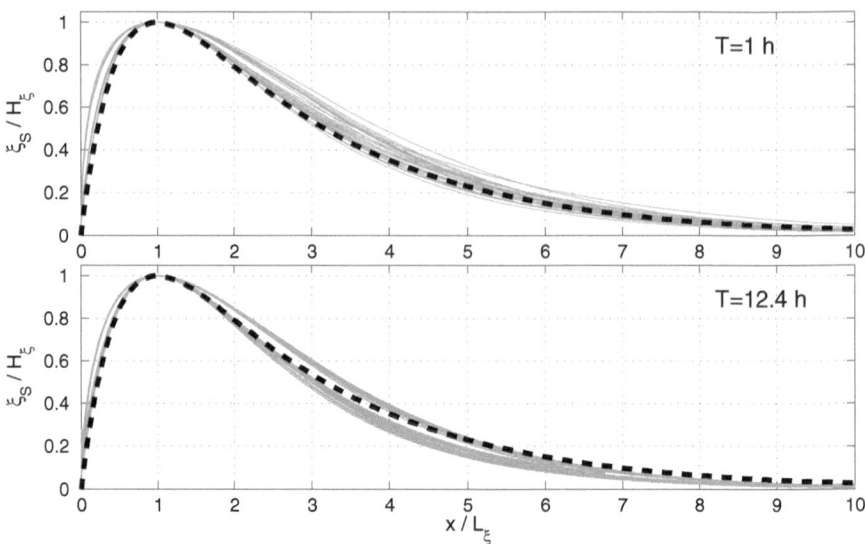

Fig. 4.6 Backwater height along a river. The backwater height is scaled by its maximal height in each scenario, while the upriver distance is scaled by the backwater accumulation distance. Gray lines: backwater curves in 27 scenarios with $T = 1$ h (top), and $T = 12.4$ h (bottom). Dashed black: analytical curve given by (2.49)–(2.51) with $\alpha = 0.22$

4.5.2   Backwater Accumulation Distance

Combining (2.50), (2.45), and (2.49), and noting that in all simulation scenarios $u_0^2 \ll c^2$, we obtain:

$$L_\xi \approx \frac{\alpha \ln(\alpha)}{\alpha - 1} \cdot \frac{c^2}{3f_0} \approx \frac{\alpha \ln(\alpha)}{\alpha - 1} \cdot \frac{h_0}{3\beta_0} \tag{4.12}$$

As expected (see Sect. 2.1), in geometrically-similar rivers, the distance L_ξ is proportional to the river's depth h_0. Furthermore, once it's been established that α is a constant, Eq. (4.12) confirms the observed pattern E, that the backwater accumulation distance is inversely proportional to the bed slope.

The same conclusion also follows from a mere suggestion that L_ξ/h_0 depends solely on a bed slope via an arbitrary function $\phi(\beta)$, should we consider a relation between the solutions of Problems I and II of the First Analytical Chapter, Sect. 2.1. Let $L_{\xi,2} = h_0 \cdot \phi(\beta)$ characterize the solution of Problem I with $T = T_2$; and $L_{\xi,3} = h_0 \cdot \phi(T_2/T_1 \cdot \beta)$—that of Problem II. Then according to (2.24), $L_{\xi,2} = T_2/T_1 \cdot L_{\xi,3}$, or $\phi(\beta) = T_2/T_1 \cdot \phi(T_2/T_1 \cdot \beta)$. The latter translates into $\phi(\beta) \sim 1/\beta$.

Extrapolating (4.12) to a case $\beta_0 \to 0$, which constitutes an endless channel with no outgoing current, results in $L_\xi \to \infty$. Indeed, as illustrated in Fig. 3.1, the maximal height of the wave set-up in such a channel would be reached infinitely far from the channel's entrance.

Substituting $\alpha = 0.22$, $h_0 = 5\,m$, and $\beta_0 = 0.11\,m/km$ into (4.12) results in $L_\xi = 6.5\,km$, while the simulations left a value of $8\,km$ for that bed slope and $T = 12.4\,h$ (a case with no flow reversal). An approximate nature of our analytical estimates stems from an assumption (2.47) used for facilitating analytical manipulations.

4.5.3   Backwater Height and Wave Period

Transformation (2.24) preserves vertical sizes, such as a backwater height. As follows from the relation between Problems I and II of the First Analytical Chapter, should the backwater height be lower for steeper bed slopes (data pattern D), then longer wave periods would also yield lower backwater heights (as confirmed by data pattern A). Apparently, for a very long period and a given wave amplitude, the backwater height at each location is a mere average of static backwater heights corresponding to water levels at the mouth at different wave phases. This average is therefore the lowest limit for the backwater height induced by a harmonic wave with a given amplitude. In Fig. 4.7, we show static backwater curves in a river with parameters $\beta_0 = 0.11$, $n = 0.04$, and $u_0 = -0.75$, set by different water levels at the mouth. These water levels were sampled from one period of a cosine function with 2.52 m amplitude. The average of the corresponding backwater curves is the river's mean stage, up to 0.31 m high, which would be set by a wave with an indefinitely long period. A shorter wave sets the river stage higher. A harmonic wave with a tidal period of 12.4 h causes an uplift up to 0.56 m, while a harmonic wave with a tsunami period of 1 h raises the river's mean stage exactly twice as high (see Fig. 4.7). Regardless of the period, the maximal uplift occurs at approximately the same distance from the mouth, at 7–8 rkm.

4.6   Two Metrics for Wave Decay with Distance Travelled

The last two response parameters (4.7) and (4.8) describe a wave intrusion in a river in a more conventional way. The presence of a wave manifests itself to a bystander as an alternating rise and fall of the river surface, of which the maximal surface rise presents a greater concern as a hazard. This is also the quantity imprinted in high water marks, and measured by field surveys. Perhaps the major relevant question is how far upriver these marks can be found, and the wave-like motion can be seen, or how fast the wave effects vanish with the distance. The "wave-like motion" is characterized by surface elevation variance $w(x)$ given by (4.7),

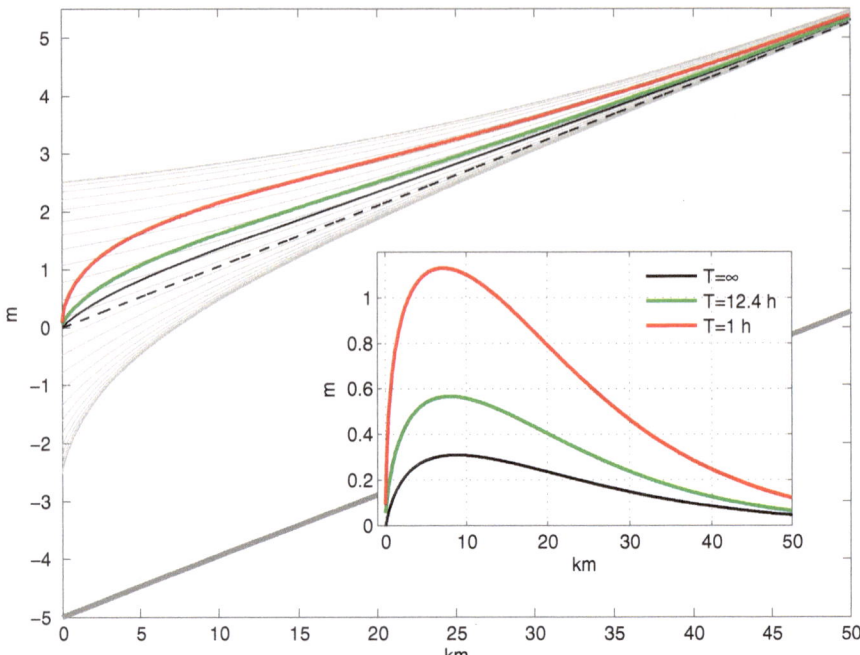

Fig. 4.7 Main panel: light gray—river's surfaces at different static water levels at $x = 0$, sampled at equal phase intervals from one period of cosine motion with $a = 2.52$ m amplitude. Black—the mean of the static surfaces; dashed black—the static river surface for the mean (undisturbed) "sea" level; thick gray—the river's bottom; color—the river's mean stages set by waves with a 2.52 m amplitude and a period $T = 1$ h (red) and $T = 12.4$ h (green). Inner panel: the river's mean stage uplift (backwater height) along the river in respond to surface level oscillations at the mouth with a 2.52 m amplitude and different periods: indefinitely slow ($T = \infty$) (black), a 1 h period (red), and a 12.4 h period (green). A backwater curve in the inner panel shows an elevation of a respectively-colored curve above the dashed-black curve, in the main panel. River parameters are $h_0 = 5$ m, $\beta_0 = 0.11$ m/km, $n = 0.04$ s $\cdot$ m$^{-1/3}$, and $u_0 = -0.75$ m/s

whereas the surface departure is characterized by its maximal instant rise at each point $\xi_{max}(x)$ given by (4.8). Note, that the surface rise has contributions from both the "wave-like motion" and the backwater. Against conventional expectations, the two metrics $w(x)$ and $\xi_{max}(x)$ express wave dissipation with a traveled distance in rather different ways.

4.6.1   Metrics I: Wave Variance

For each simulation (among 27 simulations with one wave period, and 27—with another), a natural logarithm of the variance ratio $\ln(w(0)/w(x))$ is computed as a function of x/L_ξ. The resulting curves (referred to as dissipation curves) are

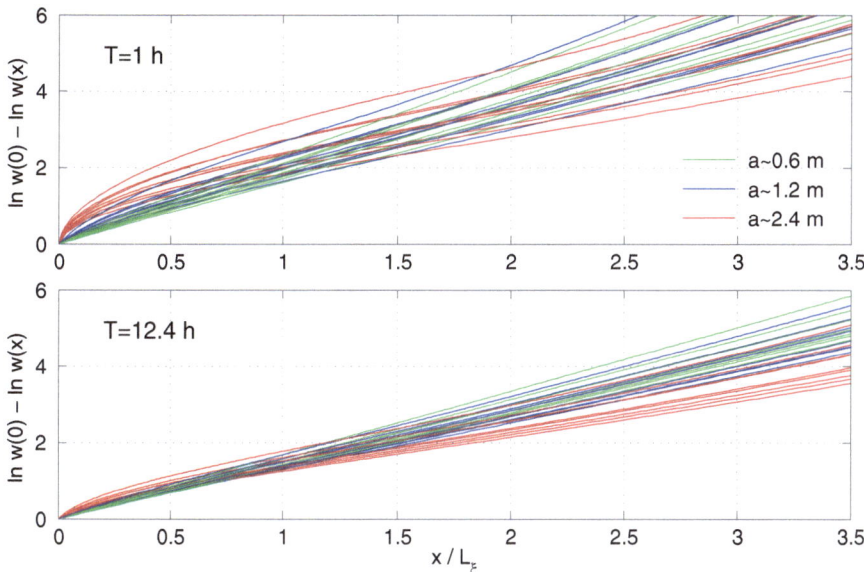

Fig. 4.8 Logarithm of wave variance ratio $w(0)/w(x)$ as a function of an upriver distance scaled by a respective backwater accumulation distance, in all scenarios with $T = 1\,\mathrm{h}$ (top), and $T = 12.4\,\mathrm{h}$ (bottom). Curves are color-coded according to the wave amplitude at $x = 0$ (the mouth)

plotted in the same axes for the same wave period, in Fig. 4.8. Should the wave decay exponentially as assumed by Eq. (2.47), then $\ln(w(0)/w(w)) = x/L_w$, and the dissipation curves in Fig. 4.8 would be straight lines with slopes $k_w = L_\xi/L_w$, where L_ξ and L_w are specific for each scenario. The factor k_w represents a rate of wave dissipation with a travelled distance, should the distance be expressed in units of L_ξ. A larger slope k_w corresponds to a shorter (relative to L_ξ) dissipation distance L_w.

The observed behavior of the computed parameters is somewhat more complicated. First, in this metrics, wave attenuation with the distance significantly deviates from an exponential law, unless the wave amplitude is small. Evaluating L_w with (2.50) and $\alpha = 0.22$ yields $L_w = 0.51L_\xi$, and $k_w \approx 2$—fairly close to the slopes of green curves representing waves with smaller amplitudes in Fig. 4.8.

To facilitate inter-comparisons, Fig. 4.9 shows local dissipation distances L_w in dimensional units (km). Local L_w values are derived from the slopes of dissipation curves in different river segments as L_ξ/k_w, in all 54 scenarios. The top plot shows L_w near the river mouth, while the bottom plot shows L_w in an upriver segment $L_\xi < x < 2L_\xi$ past the peak wave set-up. Local dissipation distances are plotted against an incoming wave amplitude. The data are color-coded according to the bed slopes. The circles and triangles indicate "tide" and "tsunami", respectively.

The immediate conclusions are that (1) the milder the bed slope, the slower a wave dissipates (that is, it dissipates over the longer distances L_w); and (2) in this metrics, "tide" dissipates slower than "tsunami". The latter can be readily explained

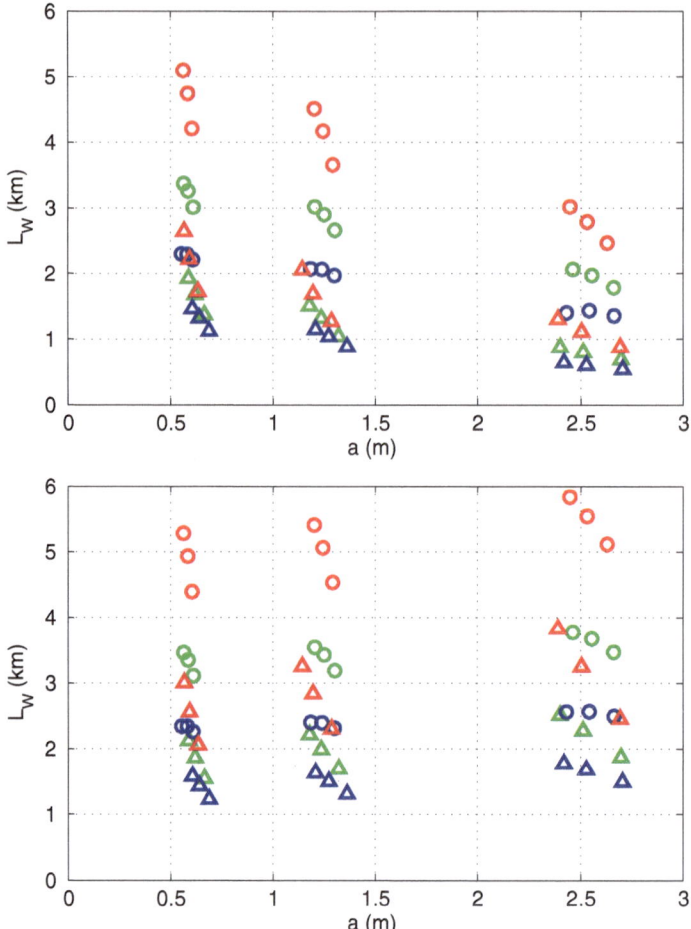

Fig. 4.9 Local dissipation distance L_w (km) in a river segment near the mouth $0 < x < L_\xi/4$ (top), and in an upriver segment $L_\xi < x < 2L_\xi$ (bottom). Colors refer to different bed slopes: 0.11 m/km (red), 0.19 m/km (green), and 0.29 m/km (blue); markers—to waves with periods 1 h (triangles) and 12.4 h (circles)

by that the shorter-period wave carries a greater flow velocity (compare lower rows of plots in Figs. 4.1 and 4.2) and therefore meets a greater friction. Other dissipation patterns vary along the river. Dissipation rate of large waves near the mouth is greater than that of small waves (Fig. 4.9, top). This wave behavior makes perfect sense, given that the friction is quadratic in flow velocity, and therefore a large wave is subjected to especially large dissipation. Farther upriver on the top of the wave set-up, however, the dissipation rates for the large waves diminish by factors of 2 to 3 (Fig. 4.9, bottom). In this segment, what's left of a large wave dissipates even slower than what's left of a small wave. This paradoxical reversal of the relative dissipation rates can be explained by backwater accumulation. Farther upriver, a large wave makes a river deeper, and in doing so facilitates its own propagation. More backwater under large waves increases the flow depth and reduces the mean flow, which results in reduced friction.

4.6.2 Metrics II: High Water Marks

Likewise, natural logarithms of the surface excursion ratio $\ln(\xi_{max}(0)/\xi_{max}(x))$ are plotted in Fig. 4.10 as functions of x/L_ξ, in the same axes for the same wave period. *The picture changes in many aspects:*

- In this metrics, wave dissipation with the travelled distance closely follows an exponential law, regardless of the wave amplitude. This is a remarkable circumstance, given that exponential dependencies belong to linear propagation mediums, which a river is not.
- The slopes of the corresponding dissipation curves are approximately the same for the waves at both periods. The high water marks left by "tide" and "tsunami" in comparable scenarios are going to be at about the same level, as we already saw in simulations in Sect. 4.4.

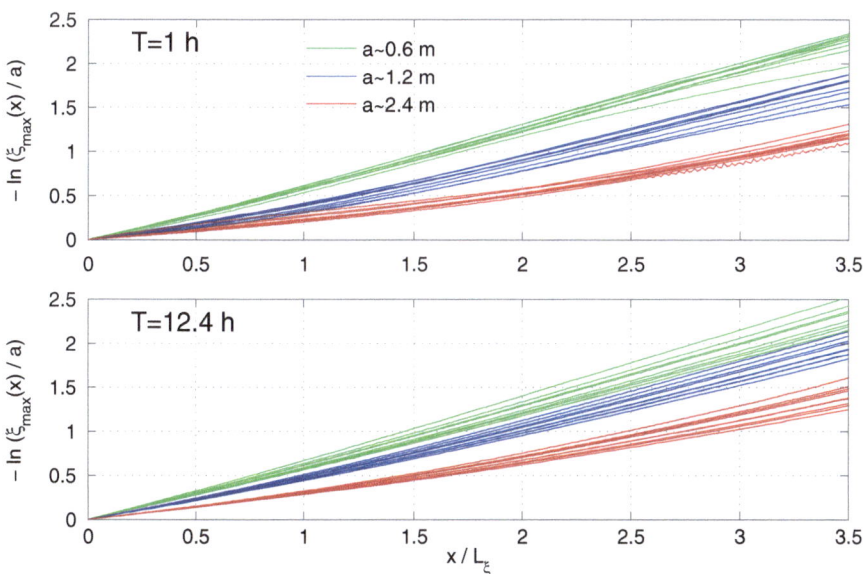

Fig. 4.10 Logarithm of surface excursion ratio $\xi_{max}(0)/\xi_{max}(x)$ as a function of an upriver distance scaled by a respective backwater accumulation distance, in all scenarios with $T = 1$ h (top), and $T = 12.4$ h (bottom). Curves are color-coded according to the wave amplitude the mouth

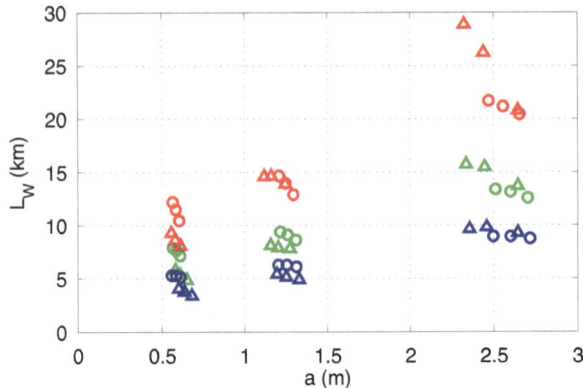

Fig. 4.11 Metrics II based e-folding dissipation distance. Colors refer to different bed slopes: 0.11 m/km (red), 0.19 m/km (green), and 0.29 m/km (blue); markers—to waves with periods 1 h (triangles) and 12.4 h (circles)

- The dissipation curves clearly group together according to the wave amplitude at the mouth. In this metrics, which combines both the wave variance and the wave set-up, a larger wave always has a lower dissipation rate.

Figure 4.11 shows metrics-II-based dissipation distances L_w in dimensional units (km) in all 54 scenarios plotted against an incoming wave amplitude. The dissipation rate/distance was calculated in a river segment $0 < x < 3L_\xi$, and assumed constant river-long. The data are color-coded according to the bed slopes. The circles and triangles indicate "tide" and "tsunami", respectively.

The metrics-II-based dissipation distance seems to primarily depend on two factors: a bed slope and an intruding wave amplitude. As in metrics I above, the milder the surface slope, the slower the wave dissipates, and the farther upriver it penetrates. Field data fully support this proposition (Tanaka et al. 2014). Counterintuitively, a wave also dissipates slower, when its amplitude is greater. In each river, the dissipation distance doubles when the wave amplitude changes from about 0.6 to about 2.5 m, even though the larger wave experiences more friction. Dissipation distances for "tide" and "tsunami" are close together. The metrics II e-folding dissipation distance starts at about $1.5L_\xi$ for small waves, and increases for larger waves, with little to no dependance on the wave period. Dissipation distances[4] in metrics II (up to 30 km, in our scenarios) are much longer than those in metrics I (up to 12 km)—quite an effect of including the wave set-up into considerations!

An estuarine tidal scientist once informed the author of this book that a tsunami's intrusion distance into a river would be much smaller than that of tide, because the higher frequency (semi-diurnal) tidal constitutes propagating from the ocean

---

[4]Note that L_w is an e-folding penetration distance for the wave intensity. Such a distance for the wave height or amplitude is $2L_w$.

typically diminish upriver faster than the longer frequency (diurnal) constituents. Then a tsunami, as just a "short tide" (that is, a hypothetical constituent with a short period), would diminish upriver even faster. However, a tsunami is more than just a "short tide", nor does a tidal constituent in a non-linear riverine environment constitute a valid physical entity. At the very least, a tsunami is a "short tide" and its backwater. This backwater is much higher than that of a "long" tide. Backwater height compensates for the smaller range of surface oscillations, resulting in approximately the same maximal surface elevation at all wave periods.

4.7 Back to the Real World

Real-world rivers have more complex morphologies than river models in our simulations. The river's "behavioral" parameters are likely to be affected by at least two more factors—the channel's convergence, and the channel's meandering. In particular, in a convergent river having the same bed elevation, discharge, and upstream width/depth as a non-convergent counterpart, the maximal wave set-up is higher and occurs father upstream (Tolkova et al. 2015). However, the same physical laws remain in play, and we don't expect the observed wave patterns to change dramatically. The quintessence of these patterns follows:

- the backwater accumulation distance L_ξ (defined as a distance from the mouth to the point of the maximal wave set-up)

  - provides a spatial scale for all wave effects,
  - is roughly independent of the wave amplitude and period,
  - depends primarily on the bed/surface slope, and is greater when the slope is milder;

- whereas the backwater height

  - depends primarily on the wave amplitude and period,
  - is roughly proportional to the square of the wave amplitude,
  - is about twice as high for a tsunami than for tide with the same amplitude; and

- the e-folding dissipation distance corresponding to high water marks

  - is of order of L_ξ for small waves,
  - gets longer for larger waves,
  - is roughly independent of the wave period.

Limited field observations support some of the above conclusions. Analyzing available water level measurements and field survey data of the 2010 Chilean and 2011 Tohoku tsunamis along six rivers in Japan, and of the 2004 Indian Ocean tsunami along three rivers in Banda Aceh, Indonesia, Tanaka et al. (2014) found that a tsunami's intrusion distance in a river correlates to the riverbed slope, being longer in rivers with milder slopes. Tanaka et al. (2014) approximated the intrusion distance with an exponentially decreasing function of the slope. However, given an uncertainty of determining attenuation coefficients and intrusion distances in a limited range of slopes away from zero, it might be difficult to reliably distinguish an exponential dependance $e^{-\beta}$ and an inverse dependance $1/\beta$, emerged in our investigation.

The relations found in this book allow us to use everyday tidal observations as an aid for predicting the penetration distance and upriver impacts of a large tsunami. For instance, in Sect. 3.2.1, analysis of the regular tidal wave set-up was successfully applied to hindcast the temporal river stage elevation during the 2011 Tohoku tsunami in the Yoshida River. The backwater accumulation distance, which can possibly be found from tidal data, provides a reference for the tsunami intrusion distance. Note, however, that the conclusions obtained herein describe an established wave regime in a river due to harmonic forcing at the mouth. In practice, it might take several tsunami waves (the shorter the period, the more waves it takes) to elevate the river's stage to its estimated position.

References

Lynett, P., Gately K., Wilson R., Montoya L., Arcas D., Aytore B., et al. (2017). Inter-model analysis of tsunami-induced coastal currents. *Ocean Modelling, 114*, 14–32. http://dx.doi.org/10.1016/j.ocemod.2017.04.003. ISSN1463-5003.

Tanaka, H., Kayane, K., Adityawan, M. B., Roh, M., & Farid, M. (2014). Study on the relation of river morphology and tsunami propagation in rivers. *Ocean Dynamics, 64*(9), 1319–1332. https://doi.org/10.1007/s10236-014-0749-y.

Tolkova, E. (2014). Land-water boundary treatment for a tsunami model with dimensional splitting. *Pure and Applied Geophysics, 171*(9), 2289–2314. https://doi.org/10.1007/s00024-014-0825-8.

Tolkova, E., Tanaka, H., & Roh, M. (2015). Tsunami observations in rivers from a perspective of tsunami interaction with tide and riverine flow. *Pure and Applied Geophysics, 172*(3–4), 953–968. https://doi.org/10.1007/s00024-014-1017-2.

Chapter 5
Second Analytical Chapter: Ascending with the Wave

Highlights Characteristics, Wave-Locked Slope (WLS), and a new form of the SWE. Analogy between a steady flow and a wave propagating into a channel. WLS balances friction. Interplay of friction and channel shape variations in amplifying/attenuating the wave. When does a wave crest keep its elevation along a river? Excursus into analytical treatment of river tides. Conventional assumptions, which we will not make. Inclinations of the High Water and Low Water trajectories. Effect of tidal WLS on a small tsunami ascending a river.

5.1 Wave-Locked Slope, and a New Form of the Equations

Following the First Analytical Chapter, patterns of long wave propagation in rivers will be deduced here using the cross-sectionally averaged Shallow-Water Equations (SWE):

$$h_t + (hu)_x + hu \cdot b_x/b = 0 \quad OR \quad A_t + (Au)_x = 0, \tag{5.1}$$

$$u_t + uu_x + gh_x = gd_x + f \quad OR \quad u_t + uu_x + g\eta_x = f, \tag{5.2}$$

where subscripts denote partial derivatives, the x-axis is oriented upriver with $x = 0$ at the river mouth; u is the cross-sectionally averaged x-component of the fluid velocity; b is the river breadth considered time-invariant; A is the cross-sectional area of the flow; h is the average flow depth; d is the average river bed elevation counted *down* from the reference level; η is the free surface elevation *above* the reference level; g is the acceleration due to gravity; and f is the friction term.

Given the original bed elevation $\tilde{d}(x, y)$, velocity $\tilde{u}(t, x, y)$, and flow depth $\tilde{h}(t, x, y) = \tilde{d} + \eta$ defined in the along-river x and cross-river y coordinates, the above averages are:

$$A = \int_{y_1}^{y_2} \tilde{h} \cdot dy, \quad u = \frac{1}{A} \int_{y_1}^{y_2} \tilde{u} \cdot \tilde{h} \cdot dy = \frac{Q}{A}, \quad d = \frac{1}{b} \int_{y_1}^{y_2} \tilde{d} \cdot dy, \tag{5.3}$$

© The Author(s) 2018
E. Tolkova, *Tsunami Propagation in Tidal Rivers*, SpringerBriefs
in Earth Sciences, https://doi.org/10.1007/978-3-319-73287-9_5

$$h = \frac{A}{b} = d + \eta, \tag{5.4}$$

where $y_1(x)$ and $y_2(x)$ are the left and right shoreline coordinates, $b = y_2 - y_1$, Q is the discharge rate. Also, as discussed in Chap. 2,

$$f = -\tilde{C}_d \cdot \frac{u|u|}{h} = -\frac{u|u|}{L_f}, \quad L_f = \frac{h}{\tilde{C}_d}, \tag{5.5}$$

where an effective drag coefficient $\tilde{C}_d \geq C_d$ is a product of a true drag C_d and the factors depending on the flow non-uniformity. The less uniform the flow and the bed elevation across the river, the greater the effective drag. The length L_f represents a spatial scale of friction from a perspective of a fluid parcel.

The cross-sectionally averaged SWE (5.1)–(5.2) are derived from the original SWE under the assumptions that the flow occupies a straight channel, and the flow velocity is fairly uniform across it, so that the momentum correction coefficient can be set to unity. The channel width is considered time-invariant, and the water surface is considered horizontal across the river. No other assumptions or approximations while manipulating the SWE (5.1)–(5.2) will be made hereafter.

Using the method of characteristics (Stoker 1957; Courant and Hilbert 1962; Whitham 1974), the SWE can be re-written in terms of Riemann invariants ζ^r and ζ^l

$$\zeta^r = u + 2c, \quad \zeta^l = u - 2c, \quad c = \sqrt{gh} \tag{5.6}$$

as a system of hyperbolic equations

$$\zeta^r_t + (u + c)\zeta^r_x = gd_x + f - uc \cdot b_x/b, \tag{5.7}$$

$$\zeta^l_t + (u - c)\zeta^l_x = gd_x + f + uc \cdot b_x/b. \tag{5.8}$$

As follows from (5.6),

$$\zeta^l_t = u_t - ch_t/h, \quad \zeta^l_x = u_x - ch_x/h, \tag{5.9}$$

$$\zeta^r_t = u_t + ch_t/h, \quad \zeta^r_x = u_x + ch_x/h. \tag{5.10}$$

It is straightforward to check that the equation set (5.7)–(5.8) is equivalent to the set (5.1)–(5.2). Subtracting (5.8) from (5.7) simplifies to the continuity equation (5.1), while adding them simplifies to (5.2). A local flow state is equivalently determined by the state variables (u, h) or by the Riemann invariants (ζ^r, ζ^l). Hereafter, the term "local" refers to a specific position in space and time. The Riemann invariants are said to "propagate" along corresponding characteristics: the right-going invariant ζ^r "propagates" upstream along a space-time trajectory $dx/dt = c + u$, whereas the left-going invariant ζ^l "propagates" downstream along a trajectory $dx/dt = u - c$ (presumingly, $|u| < c$). Should, for instance, ζ^l be constant at some point A, then no wave motion upstream of A contributes to the flow state downstream of A. The condition $\zeta^l = const$ at the river mouth implies that no fraction of the incoming wave is reflected back to the sea.

Tsunami observations in rivers brought about a physical quantity termed Wave-Locked Slope (WLS) (Tolkova et al. 2015). The local WLS for a right-going wave is:

$$\eta_s = \eta_x + \eta_t/(c+u) \tag{5.11}$$

where the subscript s represents spatial differentiation along the right-going characteristic:

$$d/ds = \partial/\partial x + 1/(c+u) \cdot \partial/\partial t. \tag{5.12}$$

An observer locked with a specific phase in a right-going wave, while traveling a horizontal distance Δx through time $\Delta t = \Delta x/(u+c)$, moves vertically by $\Delta y = \eta(x+\Delta x, t+\Delta x/(u+c)) - \eta(x,t) = \eta_s(x,t) \cdot \Delta x$. Therefore, η_s represents a slope of the free surface seen by this observer. Apparently, for a steady flow, WLS coincides with the surface slope η_x. Also, for a time-invariant quantity, $d/ds = \partial/\partial x$, so for instance,

$$A_s = b_s h + h_s b = b_x h + (d_x + \eta_s)b, \tag{5.13}$$

and

$$Q_s = (uA)_s = u_s A + u(b_x h + d_x b + \eta_s b) = uA\left(\frac{u_s}{u} + \frac{b_x}{b} + \frac{d_x}{d} + \frac{\eta_s}{h}\right). \tag{5.14}$$

Next, we attempt to re-write the SWE to follow the motion along the right-going characteristic. Substituting (5.10) into (5.7), replacing h_t with η_t, h_x with $d_x + \eta_x$, and regrouping terms yields:

$$((u+c)u_x + u_t) + \frac{c}{h}((u+c)\eta_x + \eta_t) = gd_x + f - uc\frac{b_x}{b} - (u+c)c\frac{d_x}{h}$$

Recognizing derivatives with respect to s in the left part, and using the equality $c^2 = gh$ simplifies the last equation to

$$(1 + u/c)(cu_s + g\eta_s) = f - uc(b_x/b + d_x/h) \tag{5.15}$$

After another regrouping of the terms, Eq. (5.15) can be expressed as

$$uu_s + g\eta_s = f - cu(u_s/u + b_x/b + d_x/h + \eta_s/h), \tag{5.16}$$

or given (5.14), as

$$uu_s + g\eta_s = f - cQ_s/A. \tag{5.17}$$

The structure of the resulting equation resembles that of the momentum equation. Equation (5.17) will replace the momentum equation (5.2) in the new SWE set.

To replace the continuity equation, a few more manipulations follow. For a time-invariant breadth

$$(uA)_x = (uA)_s - \frac{b}{c+u}(uh)_t,$$ (5.18)

and

$$b \cdot h_t - \frac{b}{c+u}(uh)_t = -\frac{bh}{c+u}(u_t - ch_t/h) = -\frac{A}{c+u}\zeta_t^l,$$ (5.19)

so

$$A_t = -\frac{A}{c+u}\zeta_t^l + \frac{b}{c+u}(uh)_t.$$ (5.20)

Adding up (5.18) and (5.20) to form the continuity equation yields:

$$-\frac{A}{c+u}\zeta_t^l + Q_s = 0.$$ (5.21)

Equations (5.17) and (5.21), with ζ^l defined by (5.6), represent another form of the SWE. As such, the new equations are exact under the shallow-water approximation.

5.2 Unidirectional Wave and Steady Flow

In a simple unidirectional wave traveling right, no variance propagates along the left-going characteristics: $\zeta_t^l = 0$. Then the newly derived SWE set becomes:

$$Q_s = 0, \quad uu_s + g\eta_s = f.$$ (5.22)

For comparison, the good old SWE for a steady flow simplify to:

$$Q_x = 0, \quad uu_x + g\eta_x = f.$$ (5.23)

We have found that equations describing a unidirectional wave in a channel with negligible reflection are identical to those describing a steady flow, but map onto s instead of x. In particular, the "old" continuity equation (5.23)-left states, that the discharge rate Q is the same at every location along the river. The "new" equation (5.22)-left states, that an observer moving with the wave would measure the same discharge rate Q at subsequent locations and times. Next, WLS relates to local friction under a unidirectional wave in a manner similar to that in which the stationary surface slope relates to friction under the steady flow. Steady slope balances friction upon water particles traveling with the flow. WLS balances dynamically changing friction upon a sequence of particles transferring the wave momentum. This connection between WLS and dissipation was originally discovered in observations of small-amplitude tsunamis propagating in tidal rivers and systematically modulated according to the tidal WLSs ascended by the tsunami waves (Tolkova et al. 2015; Tolkova 2016).

In the real world, however, an upriver-going wave trajectory starts at the mouth with the discharge rate prescribed by both the local sea conditions and the riverine discharge, and ends upstream past the wave penetration area, where the discharge is set by the river only. Therefore, the discharge rate along the wave path is bound to vary. A condition $Q_s = 0$ cannot take place in a frictional medium, since this condition implies that the wave never dissipates. Consequently, there is always a reflected wave (defined as a variance propagating along the left-going characteristics toward the river mouth) in a frictional channel.

However, friction not only contributes toward reflection, but also damps it. Arguing that friction dominates the momentum balance, LeBlond (1978) concluded that the SWE in riverine environments can be approximated with a parabolic equation permitting propagation only up the river. Since then, neglecting reflection has become a customary practice in estuarine studies, where a solution for either the tidal elevation or velocity is commonly sought in a form of a unidirectional wave decaying with an upriver distance (Godin 1985; Horrevoets et al. 2004; Jay 1991; Savenije et al. 2008, as a few examples). As follows from (5.17), a condition permitting to neglect the reflected wave reads

$$c|Q_s|/A \ll |f|, \tag{5.24}$$

or given (5.21), it reads

$$|\zeta_t^l| \ll |f|. \tag{5.25}$$

The latter form re-iterates the premise behind neglecting the reflected wave, stating that the wave intrusion is affected primarily by friction. Thereby, even in a realistic frictional channel, where the first equation in (5.22) does not hold, the second equation might still be approximately correct. Routinely neglecting variation in u^2 next to variation in $g\eta$ for well-subcritical flows,[1] the latter equation approximates WLS with local friction:

$$g\eta_s = f. \tag{5.26}$$

WLS balances friction under a wide range of natural conditions—similar to a hydraulic slope which balances friction in a stationary flow. At the same time, there is a difference between a steady slope and WLS: the steady slope creates a

---

[1]In particular, in tidal rivers studies, the convective term uu_x is often dismissed as negligible compared to either the friction term or $g\eta_x$ (LeBlond 1978; Godin 1999).

pressure gradient which compensates friction and supports the flow, whereas WLS only partially converts into pressure, and therefore does not prevent the wave from dissipating.

The WLS concept might allow measuring frictional properties of river beds or river currents using water level records along rivers.

5.3 When Channel Convergence Balances Friction

In natural rivers, condition (5.24) gets broken when reflection (and consequent variation in the water transport along a characteristic) is caused by a varying channel's shape rather than by friction. In a 1-D river model, a channel's shape is represented by the average bed elevation and by the channel's cross-sectional area occupied by the flow. The channel is said to converge when its cross-section gradually narrows. An "ideal estuary"—a 1-D river model commonly adopted for analytical studies of tides in rivers—has an exponentially convergent channel with a flat bottom and an exponentially narrowing breadth (consequently, the freshwater discharge has to be zero).

Physically, a discharge carried upriver with a particular wave phase can only decrease in magnitude, due to both wave dissipation and reflection. The narrowing of the channel can increase the flow velocity, but not the cross-river integrated discharge carried by the wave. On the contrary, the discharge can only decrease, due to likely reflection from the narrows. Thus we expect the discharge to vary gradually along the characteristic, from its value at the mouth to a given riverine discharge at $x = \infty$, with $Q_s > 0$ during outflow, and $Q_s < 0$ on inflow (Fig. 5.1). Closer to the mouth, the friction is directed upriver (positive) on outflow as well, and negative on inflow, so the term cQ_s/A in the new "momentum" equation (5.17) reduces the

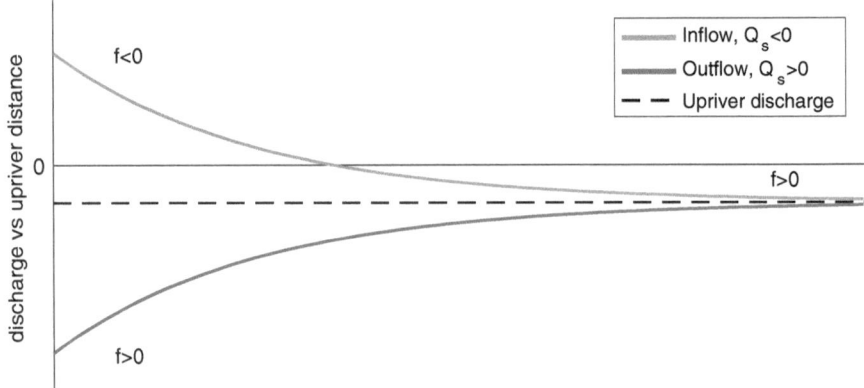

Fig. 5.1 Discharge variation with an upriver distance, during inflow and outflow phases

dissipative action of friction upon the upriver-propagating wave. As follows from the latter equation, the condition

$$f - c(uA)_s/A = 0 \tag{5.27}$$

describes the balance between the friction and the channel convergence. According to (5.17), the quantity $(g\eta + u^2/2)$ remains constant along such a characteristic path, where the balance (5.27) holds. If the term uu_s can be neglected compared to $g\eta_s$, then (5.27) becomes a condition for a zero WLS: the wave phase propagates at a constant elevation η.

Let a tsunami be the only flow component. Consider a convergent channel with a negligible freshwater current, such as an "ideal estuary" or a narrow inlet—a simplified system which does not have a hydraulic surface gradient and a current of its own, and where η and u belong entirely to an intruding wave. Then $(uA)_s/A = u_s + uA_s/A$, with $u_s \sim u/L_w$ and $A_s/A \sim 1/L_A$, where L_A is an e-fold length of the channel cross-section under a particular wave phase, and L_w is the wave dissipation/amplification scale.[2] Unlike u_x, u_s does not explicitly depend on the wavelength λ, though L_w might depend on it. For instance, for a wave propagating in an ideal frictionless uniform channel, $u_x \sim u/\lambda$, but $u_s = 0$ and $L_w = \infty$, since the wave in such a channel does not dissipate. Differently from friction-dominated channels considered in Sect. 5.2, we now consider channels with

$$L_w \gg L_A, \tag{5.28}$$

that is, channels which converge faster than the intruding wave dissipates or amplifies. Then $(uA)_s/A \approx uA_s/A$. After also neglecting uu_s next to $g\eta_s$, (5.17) simplifies to

$$g\eta_s \approx f - cuA_s/A = cu\left(\frac{1}{L_A} - \tilde{C}_d\frac{|u|}{ch}\right), \tag{5.29}$$

where $1/L_A = -A_s/A$, and friction is expanded as $f = -\tilde{C}_d u|u|/h$.

Using (5.13) in a flat-bottom channel, $A_s/A = b_x/b + \eta_s/h$, where the first term on the right side is $O(1/L_A)$, and the second term is $(\eta/h) \cdot O(1/L_w)$. Hence, under the assumption (5.28), the second term can be neglected. Then $L_A = -b/b_x$, which implies that L_A is determined only by the channel's geometry, and does not depend on the wave phase anymore. In an ideal estuary, L_A is a constant.

---

[2]Since our analysis is based on following the wave, we do not observe the wavelength (as seen by an observer who takes an instant picture of the river profile) nor the wave period (as seen by an observer who records the surface motion at a given point). Instead, we follow the space-time trajectory where the fluid state changes the least, and we can only observe how the wave dissipates or amplifies. Hence our only spatial scale is the wave dissipation distance L_w, and $d/ds = O(1/L_w)$.

Equation (5.29) defines an inclination of an upriver path traveled by a particular wave phase. If this slope is positive under a wave crest ($\eta > 0$, $u > 0$), then the height of high water increases upriver. If this slope is positive under a wave trough ($\eta < 0$, $u < 0$), then the trough depth decreases upriver. If on a particular wave trajectory the friction is balanced by the convergence, then the corresponding wave element propagates at a constant elevation, that is, along a horizontal path.

So, a condition that a wave crest propagates at a constant elevation reads:

$$\frac{1}{L_A} - \tilde{C}_d \frac{u}{ch} = 0, \quad or \quad \frac{u}{c} = \frac{L_f}{L_A}, \tag{5.30}$$

with L_f determined by (5.5). This is valid for well sub-critical flows in channels with $L_A \ll L_w$, after the omissions mentioned. Crests lower than prescribed by (5.30) will grow in height due to funneling prevailing over friction, while the higher crests will diminish due to friction prevailing over funneling. Thereby, (5.30) defines a wave elevation which becomes established in an idealized convergent estuary with the distance from the mouth. However, the condition (5.30) does not imply that the crest-to-trough wave span remains constant.

For some estimate of the scales involved, using Manning's $\tilde{C}_d \approx C_d = gn^2/h^{1/3}$ with $h = 5$ m and $n = 0.035$ s · m$^{-1/3}$ yields $L_f \sim 0.7$ km. Assume that $L_A = 3$ km. Then the established wave elevation can be estimated as $h \cdot u/c \approx hL_f/L_A \approx 1.2$ m.

5.4 Excursus into Analytical Treatment of Estuarine Tides

Condition (5.30)-left resembles Savenije's (2008) condition for a constant tidal amplitude in an ideal estuary, with the difference being that Savenije's condition uses the mean flow depth $\bar{h}$ and the velocity u at high water ($u = v \sin \epsilon$ in Savenije's (2012) notation, determined as the velocity amplitude times sine of the phase lag). The latter condition had been obtained by manipulating the SWE under a number of assumptions and simplifications traditional in tidal studies. Savenije et al. (2008) compiled a list of conventional assumptions made in Savenije (1992, 1993, 2001) to facilitate analytical treatment of tides in rivers. These assumptions, where 1–3 are considered the basic assumptions, and the last one—also very common, to the extent of being taken for granted—is added by us, are:

1. prescribed channel shape, exponentially narrowing as $A(x) = A_0 e^{-x/L_A}$, where A is the cross-sectional area, and L_A is a constant;
2. small tidal amplitude to depth ratio;
3. negligible freshwater discharge;
4. small Froude number (follows from assumption 2);
5. tide is symmetric and expressed by a single harmonic;
6. small tidal damping;
7. constant phase lag and wave celerity along the river;

8. small depth to width ratio and steep banks;
9. well-mixed salt intrusion;
10. momentum correction coefficient set to unity.

Assumptions made by other authors, as also pointed out by Savenije (1998), go beyond the listed ones and typically also include neglecting the advective acceleration in the momentum equation (in which case, the choice of the momentum correction coefficient becomes irrelevant), and/or linearizing the friction term, and/or assuming a zero bottom slope. Based on assumptions 5–6, the solution functions are commonly prescribed,[3] and expressed by harmonics with exponentially decaying amplitudes. These prescribed solutions are then substituted into truncated SWE to constrain the parameters (such as a wave number or a decay rate) embedded in the solutions.

Of all these assumptions, only 8–10 in the above list are used in this chapter, allowing for using the original SWE for a single fluid fraction and for neglecting the temporal variations of the river breadth. Additionally, assumptions 2, 3, and 6 were used for a particular transition from a more general condition (5.27) for the friction-convergence balance to a simplified condition (5.30).

As follows from our solution, the latter conditions apply along each characteristic path, but do not apply to the entire waveform. A path of High Water (HW) or Low Water (LW) approximately coincides with the characteristics. If we assume that in our simplified condition (5.30), the HW and LW velocities are equal in magnitude (as implied by assumption 5), and use a mean depth $\overline{h}$ instead of an actual flow depth (as permitted by assumption 2), then (5.30) can indeed be met on both HW and LW trajectories, since all the quantities involved would be the same at both tidal phases. Then both the HW elevation and the LW elevation would remain constant, and so would the tidal range, as predicted by Savenije (2008, 2012).

Under a more accurate approach, however, a tidal range in an ideal estuary is unlikely to be constant. Let the friction-convergence balance take place under High Water, which path therefore goes at a constant elevation. Then (5.30) yields:

$$\tilde{C}_d = (c^{HW} h^{HW})/(L_A u^{HW}).\tag{5.31}$$

Under the tidal trough, denoting $u = -u^{LW} < 0$ because of the outflow, and assuming the same drag $\tilde{C}_d$ for simplicity, the right part of (5.29) can be written as

$$-c^{LW} u^{LW}\left(\frac{1}{L_A} - C_d \frac{u^{LW}}{c^{LW} h^{LW}}\right) = \frac{u^{LW} c^{LW}}{L_A}\left(\frac{c^{HW}}{c^{LW}} \cdot \frac{h^{HW}}{h^{LW}} \cdot \frac{u^{LW}}{u^{HW}} - 1\right)\tag{5.32}$$

---

[3] An occurrence of prescribing unknown functions in this book is our assumption (2.47), used to facilitate wave set-up estimates. In this case, our assumption applies to mean second powers of the state variables, and does not impose a particular shape of the oscillation. Analytically computed wave set-up is verified with numerical experiments.

where both u^{HW} and u^{LW} are positive. Expression (5.32) evaluates an inclination of the LW trajectory, which orientation (up or down) is determined by the sign of the expression in brackets. Following Savenije (2012), one can assume that $u^{LW} = u^{HW}$. Then the expression in brackets simplifies to

$$(h^{HW}/h^{LW})^{3/2} - 1 > 0, \tag{5.33}$$

being positive because of the greater flow depth under HW. Alternatively, following Jay (1991) who prescribed a harmonic law to the tidal transport, one can assume that the discharge has the same value at HW and LW: $h^{LW}u^{LW} = h^{HW}u^{HW}$. Then the expression in brackets simplifies to

$$(h^{HW}/h^{LW})^{5/2} - 1 > 0. \tag{5.34}$$

In either case, WLS for the LW comes out positive. Therefore, when the surface elevation under the tidal crest (HW) remains constant, the elevation under the tidal trough (LW) climbs up, and consequently, the tidal range and amplitude reduce upriver.

In the real-world tidal records, taken in rivers where the channel shapes are more irregular than assumed in any theory, the described tendency of the HW path to be more horizontal, and the LW path to be a steep climb is nevertheless commonly observed (see Fig. 5.2 for an example). As referencing tide gages to other-than-local vertical datums is rare in the most of the world, the aforementioned tendency had remained hidden, until observations of the tsunami modulation by tide in rivers brought this tendency to light.

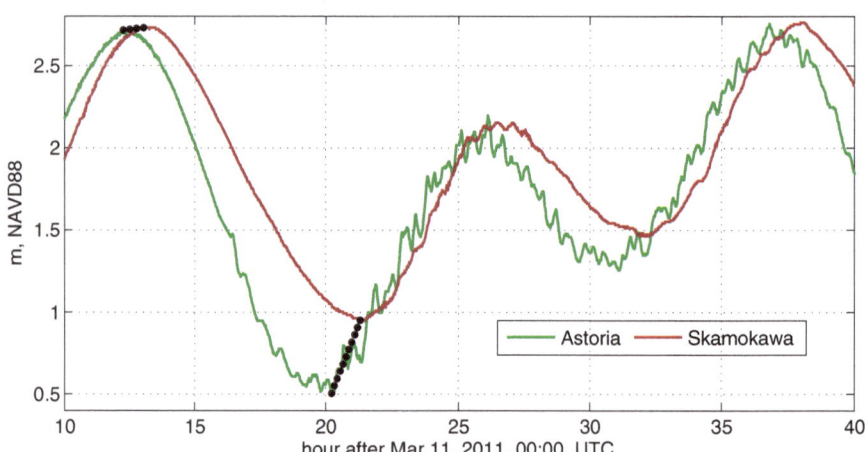

Fig. 5.2 Tidal records with traces of the 2011 Tohoku tsunami at Astoria (29 rkm) and Skamokawa (54 rkm), delimiting the convergent part of the Columbia river estuary (see Fig. 1.5 for the map). Black dotted lines show a nearly zero elevation gain in the highest water path, and a steep climb in the lowest water path

5.5 Effect of Tidal WLS on a Small Tsunami Ascending a River

Consider a small tsunami propagating with tide and riverine current in a channel, where the wave reflection is negligible. The total flow is comprised of tide-and-current (V, χ) and tsunami (v, ξ) components. Neglecting reflection, the modified momentum equation (5.22)-right expressed in terms of these two flow components reads:

$$(V + v)(V + v)_s + g(\chi + \xi)_s = f. \tag{5.35}$$

For a small-amplitude tsunami, $|v| < |V|$ during most of the tidal cycle. Then $sign(V + v) = sign(V)$, and the friction term can be expanded using

$$-|V + v|(V + v) = -\frac{|V|}{V}(V + v)^2 = -|V|V - 2|V|v + V^2 \cdot O(v^2/V^2), \tag{5.36}$$

and

$$1/L_f = C_d/h = \left(1 - \xi/h_{tide} + O(\xi^2/h_{tide}^2)\right)/L_{f,tide}, \tag{5.37}$$

where $L_{f,tide}$ and h_{tide} are the friction length and the flow depth in the tidal river in the absence of the tsunami. Neglecting terms $O(\xi^2/h_{tide}^2)$ and $O(v^2/V^2)$, the friction is approximated as:

$$f \approx -|V|V \cdot (1 - \xi/h_{tide})/L_{f,tide} - 2|V|v/L_{f,tide}. \tag{5.38}$$

The distinction between tide and tsunami is based purely on their very different time scales. The tsunami represents the higher frequency component of the flow. Applying high-pass filtering to both sides of (5.35) yields:

$$g\xi_s = T_1 + T_2 + T_3, \tag{5.39}$$

$$T_1 = -2|V|v/L_{f,tide}, \quad T_2 = |V|V \cdot (\xi/h_{tide})/L_{f,tide}, \quad T_3 = -(vV)_s. \tag{5.40}$$

Let us demonstrate, that T_1 typically largely exceeds in magnitude T_2 and T_3. Firstly, a friction length estimate at the end of Sect. 5.3 produced $L_{f,tide} = 0.7\,\mathrm{km}$, whereas a dissipation length in a 5-m deep river is commonly $L_w > 10\,\mathrm{km}$. Then T_1 exceeds $T_3 \sim vV/L_w$ by a factor $L_w/L_f > 30$. Next, as provided by (5.26) for a frictionally dominated channel, friction upon the background flow (tide and current) can be estimated with a local tidal WLS χ_s as

$$f_{tide} = -|V| \cdot V/L_{f,tide} \approx g\chi_s, \tag{5.41}$$

therefore

$$|V|/L_{f,tide} \approx \sqrt{g|\chi_s|/L_{f,tide}}. \tag{5.42}$$

The tsunami flow velocity can be *roughly* estimated as in a simple wave in a uniform frictionless channel: $v = \xi \cdot \sqrt{g/h_{tide}}$. Then the ratio $|T_1/T_2|$ evaluates to $2\sqrt{h_{tide}/(|\chi_s|L_{f,tide})}$. The denominator, $|\chi_s|L_{f,tide}$, represents the surface elevation gain over the distance $L_{f,tide}$, travelled by a particular tidal phase. Typically, this gain is much less than the flow depth h_{tide}. Thereby, $|T_1/T_2| \gg 1$ as well.

After evaluating T_1 with (5.42) and omitting T_2 and T_3, (5.39) simplifies to:

$$\xi_s \approx -2v\sqrt{|\chi_s|/(gL_{f,tide})}. \tag{5.43}$$

The resulting Eq. (5.43) describes an upriver gradient of the surface elevation travelled by a particular tsunami wave phase. The elevation is counted from the hypothetical river stage set by tide and current in the absence of the tsunami. With $\chi_s = 0$, each tsunami phase travels at a constant elevation over the tidal level—the tsunami propagates without dissipation. With $\chi_s \neq 0$, (5.43) describes a trajectory, relative to tidal WLS, with a negative inclination under a wave crest ($v > 0$) and a positive inclination under a trough ($v < 0$). Therefore, both the crest and the trough diminish upriver—the tsunami dissipates.

The greater the tidal WLS χ_s, the greater the wave dissipation rate, since the friction upon the tsunami component as given by (5.43) is determined by the wave-locked slope χ_s in the co-propagating tidal segment. A tsunami wave climbing a steep WLS is subjected to greater friction, and therefore dissipates over a shorter distance, than a distance intruded by a wave over a mild WLS. As found in Sect. 5.4, steep WLSs are likely to occur during low tide, while mild WLSs are typical for high tide conditions. The latter conditions are therefore most favorable for the tsunami intrusion.

In the next chapter, a tsunami modulation by WLS in the co-propagating tide will be found in records of the 2015 Chile tsunami in rivers on the east coast of Honshu. A presence of a typical WLS pattern over a tidal cycle, to be found again in the water level measurements, will explain an earlier observed dependance of a tsunami penetration ability on a tidal phase.

References

Courant, R., & Hilbert, D. (1962). *Methods of mathematical physics* (Vol. II). Hoboken, NJ: Wiley-Interscience.

Godin G. (1985). Modification of river tides by the discharge. *Journal of Waterway, Port, Coastal, and Ocean Engineering, 111*(2), 257–274.

Godin, G. (1999). The propagation of tides up rivers with special considerations on the upper Saint Lawrence river. *Estuarine, Coastal and Shelf Science, 48*, 307–324.

Horrevoets, A. C., Savenije, H. H. G., Schuurman, J. N., & Graas, S. (2004). The influence of river discharge on tidal damping in alluvial estuaries. *Journal of Hydrology, 294*, 213–228.

Jay, D. A. (1991). Green's law revisited: Tidal long-wave propagation in channels with strong topography. *Journal of Geophysical Research, 96*(C11), 20585–20598.

LeBlond, P. H. (1978). On tidal propagation in shallow rivers. *Journal of Geophysical Research, 83*(C9), 4717–4721.

Savenije, H. H. G., Toffolon, M., Haas, J., & Veling, E. J. M. (2008). Analytical description of tidal dynamics in convergent estuaries. *Journal of Geophysical Research, 113,* C10025. https://doi.org/10.1029/2007JC004408.

Savenije, H. H. G. (1993). Determination of estuary parameters on the basis of Lagrangian analysis. *Journal of Hydraulic Engineering, 119*(5), 628–643.

Savenije, H. H. G. (1998). Analytical expression for tidal damping in alluvial estuaries. *Journal of Hydraulic Engineering, 124*(6), 615–618. https://doi.org/10.1061/(ASCE)0733-9429(1998)124:6(615).

Savenije, H. H. G. (2001). A simple analytical expression to describe tidal damping or amplification. *Journal of Hydrology, 243,* 205–215. https://doi.org/10.1016/S0022-1694(00)00414-5.

Savenije, H. H. G. (2012). *Salinity and tides in alluvial estuaries* (2nd completely revised edition). Amsterdam: Elsevier. https://salinityandtides.com.

Savenije, H. H. G., Toffolon, M., Haas, J., & Veling, E. J. M. (2008). Analytical description of tidal dynamics in convergent estuaries. *Journal of Geophysical Research, 113,* C10025. https://doi.org/10.1029/2007JC004408.

Stoker J. J. (1957). *Water waves*. New York, NY: Interscience Publishers.

Tolkova, E. (2016). Tsunami penetration in tidal rivers, with observations of the Chile 2015 tsunami in rivers in Japan. *Pure and Applied Geophysics, 173*(2), 389–409. https://doi.org/10.1007/s00024-015-1229-0.

Tolkova, E., Tanaka, H., & Roh, M. (2015). Tsunami observations in rivers from a perspective of tsunami interaction with tide and riverine flow. *Pure and Applied Geophysics, 172*(3–4), 953–968. https://doi.org/10.1007/s00024-014-1017-2

Whitham, G. B. (1974). *Linear and nonlinear waves*. Hoboken, NJ: Wiley. ISBN 0-471-94090-9.

Chapter 6
Tsunami Rides Tides

Highlights Designated evidence of tsunami modulation by tide. Metrics for an instant tsunami amplitude. Metrics for the tidal river conditions. Separating the tidal and the tsunami signals. Omnipresent correlation with Wave-Locked Slope. Methodology of admittance computations between two along-river locations. Tackling data errors. Admittance vs WLS curves for 11 station pairs. When is a river's admittance greater? Physics of admittance variations with WLS.

As has been repeatedly observed, tsunami propagation in rivers is influenced by tidal phase. Probably the first observation of this kind was made by Wilson and Torum in records of the 1964 Prince William's tsunami in the Columbia River. They noticed that farther upriver, the tsunami signal was detectable only at high tide (Wilson and Torum 1972). Kayane et al. (2011) reported a similar effect—much higher tsunami signal at high tide than during the rest of the tidal cycle—in several rivers on the coast of Honshu intruded by the 2010 Chile tsunami. High quality records of the 2011 Tohoku tsunami along the Columbia River showed that conditions for tsunami propagation differ not only at high and low tide, but also at receding and rising tide (Yeh et al. 2012; Tolkova 2013). Tolkova et al. (2015) brought all the observations together to show that tsunami modulation by tide, in a manner in which receding and low tide dissipates a tsunami the most, and high tide dissipates it the least, is a typical pattern in rivers.

An analytical approach to finding physics behind this pattern has been attempted in the previous chapter. Here, we seek to prove and quantify tidal influence on a tsunami relying upon the field data.

© The Author(s) 2018
E. Tolkova, *Tsunami Propagation in Tidal Rivers*, SpringerBriefs
in Earth Sciences, https://doi.org/10.1007/978-3-319-73287-9_6

6.1   Designated Dataset: The 2015 Chile Tsunami in Rivers in Japan

As noted in Sect. 1.4, Chilean tsunamis are frequent visitors in Honshu rivers, where these waves are met by the MLIT water level stations. Invaluable for our study, the stations are referenced to the same vertical datum (TP). On September 16, 2015, at 22:54:33 UTC, a Mw 8.3 earthquake occurred offshore central Chile, about 500 km north of the 2010 Chile earthquake. Another trans-Pacific tsunami followed. The tsunami arrived at the Honshu east coast on September, 18 at about 6:30 JST (Japanese Standard Time). It was approximately three times lower than that in 2010, and five times higher than regular long-wave ocean noise. Nevertheless, the 2015 tsunami left clear traces at two gauging stations along Mabechi and Naruse rivers, and at three stations along Old Kitakami, Yoshida, and Naka rivers, listed in Table 6.1.

Naruse, Yoshida, and Old Kitakami all flow into the Ishinomaki Bay within the Bay of Sendai (see map in Fig. 6.1). Naruse and Yoshida merge 0.9 km from the ocean, with the most downstream station at Nobiru located in the common Naruse-Yoshida channel. Upstream of the confluence, Narise is wider, shallower and steeper than Yoshida. An average bed slope within the first 15 rkm is about 18 cm/rkm in Old Kitakami, 23 cm/rkm in Yoshida, and 35 cm/rkm in Naruse. Naka and Old Kitakami are large rivers compared to Naruse, Yoshida, and Mabechi. All five rivers have fortified banks continued with jetties at their mouthes, and don't have wide estuaries or tidelands.

Further analysis will use 2-week long water level records at 12 (altogether) stations in the five rivers. The segments of these records containing the tsunami event are shown in Fig. 6.2. The measurements are sampled at a 10-min interval. The records contain the tsunami signal, as well as ever-present long-wave noise. This

Table 6.1 List of rivers, water level stations, their distances from the river mouth, and wave travel times (in 10 min increments) from the most downstream station in each river

River	Station	x (rkm)	τ (min)
Naruse	Nobiru	0.5	–
	Ono	4.18	10
Mabechi	Shin-Ohashi	1.2	–
	Ohashi (Bridge)	4.0	10
	Nobiru	0.5	–
Yoshida	Ono	4.04	10
	Kashimadai	8.99	30
Old Kitakami	Kadonowaki	1.24	–
	Omori	13.17	30
	Wabuchi	21.78	60
	Minato	1.1	–
Naka	Suifu	12.39	20
	Shimokunii	19.71	50

From Tolkova (2016)

Fig. 6.1 North to South: Mabechi, Old Kitakami, Naruse/Yoshida, and Naka rivers on the Honshu east coast, and location of gauging stations in each river. Naruse is north of Yoshida. Map courtesy of MLIT and OpenStreetMap contributors. Reproduced from Tolkova (2016)

noise likely originates with seiches in coastal waters penetrating into the rivers. As seen in the records, several hours before the tsunami arrival, the rivers experienced a high run-off, which elevated water levels at the upstream stations for up to 2 days, through up to 1+ m. With occasional disruptions from this weather event, the records display the same tsunami penetration pattern as described in Chap. 1 for other tsunami events and rivers: a tsunami or seiche, propagating atop tides at or before the lower low water, do not penetrate to upstream stations. However, the waves do reach those stations with rising or high tide.

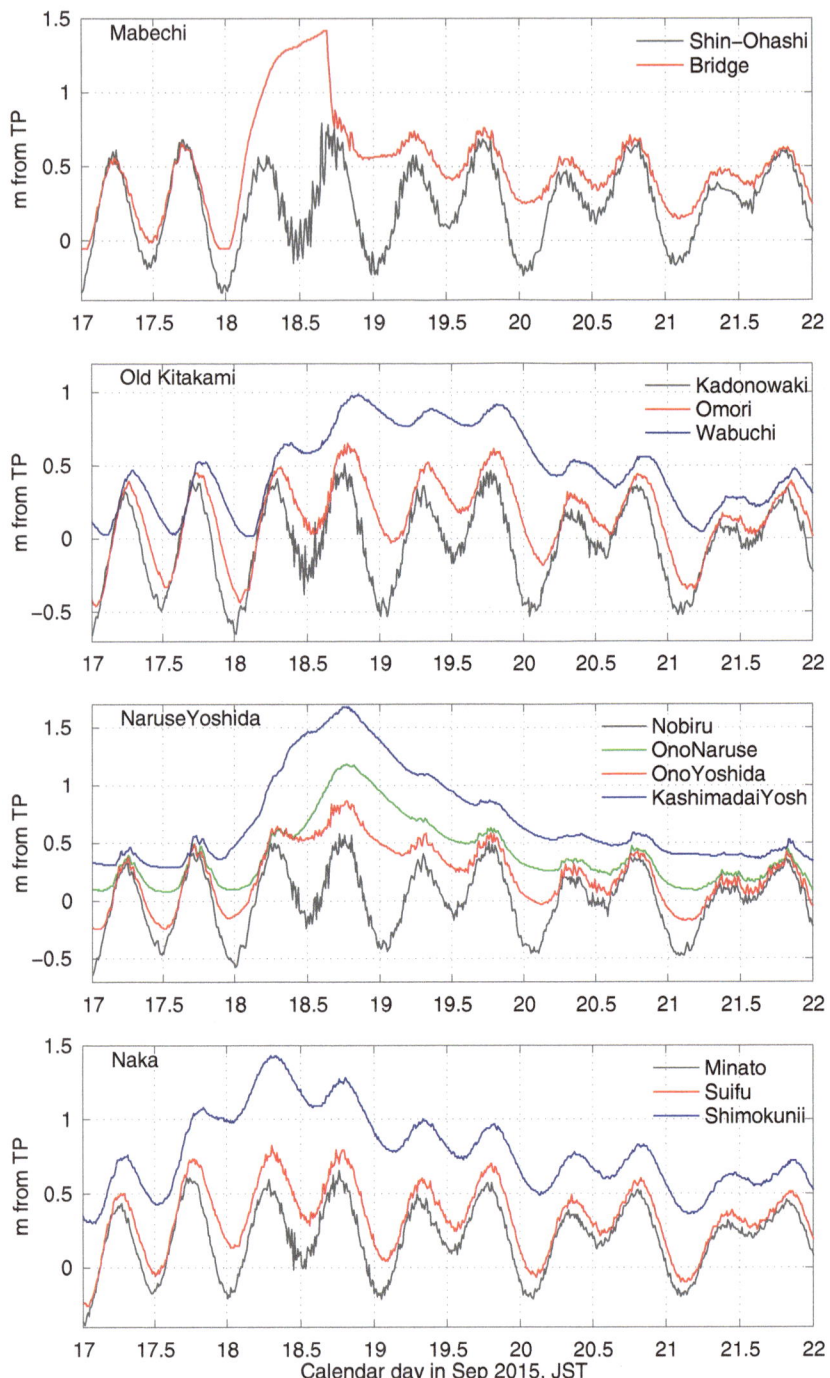

Fig. 6.2 Water elevation records with traces of the 09/2015 Chilean tsunami event along Mabechi, Old Kitakami, Yoshida and Naruse, and Naka rivers

6.2   Metrics for a Tsunami Amplitude and for the Tidal Conditions

To investigate an effect which a tidal phase has upon the tsunami's intrusion, both need to be expressed in numbers. How do we measure tsunami's attenuation/amplification at each tidal phase? How do we define and measure a state of a tidal river essential for tsunami's propagation?

Tide gage measurements can be separated into two components: the lower frequency signal $\chi(t)$ driven by an ocean tide and a riverine flow, and the higher frequency signal $\xi(t)$ seen as a disturbance on a gage record. The latter signal can be caused by a tsunami or by the long-wave ocean noise feeding normal oscillations in coastal sub-basins (seiche). The lower frequency motion, hereafter referred to as tide for brevity, sets ambient/background conditions for the shorter waves propagation in rivers.

A tsunami wave train $\xi(t)$ can be thought of as a sequence of trackable wave elements corresponding to individual water level readings. Assuming that the tide and the tsunami propagate at the same speed, each tsunami element $\xi(t)$ is locked with a tidal element $\chi(t)$ corresponding to a particular tidal phase. Both wave elements pass station A at time $t - \tau$ and arrive at station B at time t, τ being a travel time from A to B. Tsunami wave elements propagating with different tidal elements experience different "rivers", with different depth and background currents.

To quantify amplification or dissipation experienced by an individual wave element as it propagates, we assign it an amplitude. Consider a complex-valued signal

$$S(t) = \xi(t) + i\hat{\xi}(t), \tag{6.1}$$

where $i = \sqrt{-1}$, $\hat{\xi}(t)$ represents the Hilbert transform of a time series $\xi(t)$. The Hilbert transformation converts any harmonic component in $\xi(t)$, such as $\cos(\omega t)$, into its quadrature term $\sin(\omega t)$ in $\hat{\xi}(t)$. An absolute value of the complex signal $S(t)$ provides a positively-defined envelope function for $\xi(t)$, as simple examples in Fig. 6.3 demonstrate. This suggests to define an instant amplitude of a time-history $\xi(t)$ as an absolute value of the complex signal $S(t)$ (Tolkova et al. 2015). Note, that a wave element might have a non-zero amplitude even if the concurrent measurement $\xi(t)$ passes through zero.

To quantify the tidal river conditions experienced by a tsunami wave element as it propagates, we introduce a tidally-modified Wave-Locked Slope (WLS) of the river surface. An average tidally-induced WLS $\beta_{AB}(t)$ between stations A and B along a river represents an average background surface gradient traveled by a tsunami wave element from A to B. WLS can be estimated as

$$\beta_{AB}(t) = (\chi_B(t) - \chi_A(t - \tau)) / (x_B - x_A) \tag{6.2}$$

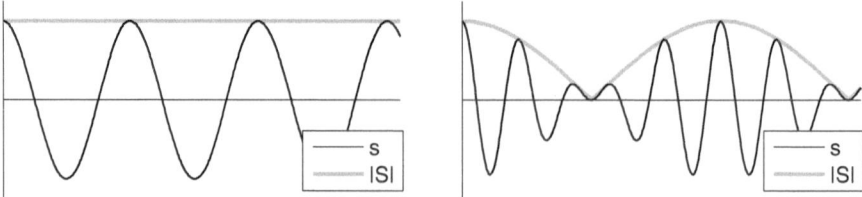

Fig. 6.3 Left: For a time series $\xi(t) = a \cdot \cos(\omega t)$, its Hilbert transform, $S(t) = a \cdot \exp(i\omega t)$, has magnitude of $|S(t)| = a$. Right: For a time series $\xi(t) = \cos(\omega_1 t) + \cos(\omega_2 t)$, its Hilbert transform, $S(t) = \exp(i\omega_1 t) + \exp(i\omega_2 t) = 2\exp(it(\omega_1 + \omega_2)/2) \cdot \cos(t(\omega_1 - \omega_2)/2)$, has magnitude of $|S(t)| = 2|\cos((\omega_1 - \omega_2)/2)|$. The absolute value of the Hilbert transform provides an envelope for the corresponding time series

where x_A and x_B are along-river distances of the locations, and τ is an A to B travel time; elevations $\chi_{A,B}$ are tidal components of water level measurements at A and B, referred to the same vertical datum. Without tidal forcing, WLS would coincide with the steady hydraulic gradient supporting the river flow. In a tidal river, WLS is continuously modified by tide. Tidal WLS will be used as a parameter defining ambient conditions for a tsunami wave element propagation.

6.3 Record's Components: Tide, Tsunami, Background Noise

To compute tidal WLS and an instant amplitude of an intruding tsunami or ocean noise, each record needs to be separated into the tidal and the shorter-wave components. The tidal component was isolated by low-pass Butterworth filtering with a 180 min cut-off period. The filter admits tides even at the upstream stations, where the tidal spectrum widens due to nonlinearity of the propagation tract. De-tiding residuals contain the tsunami and the residual background noise. At the downstream stations (see maps in Fig. 6.1 and 6.4), the recorded tsunami wave is 15–30 cm high, while the ocean noise is about 5 cm high.

Figure 6.5 displays 3-day-long segments of de-tided records at the downstream stations Nobiru in Naruse and Kadonowaki in Old Kitakami with the tsunami event on the third day, each record plotted atop a coastal record at Miyato. Both Naruse/Yoshida and Old Kitakami flow into the Ishinomaki Bay; while Miyato station is located in the Bay, at 3.8 km from the Naruse mouth. The residual background noise is practically the same at all three stations, which suggests that the noise is caused by normal oscillations (seiche) at a scale of the Ishinomaki Bay or the surrounding Bay of Sendai. Therefore, both the tsunami and the background noise in the de-tiding residuals represent waves intruding from the ocean—a condition essential for our study.

Fig. 6.4 MLIT gauging stations (orange triangles) in Ishinomaki Bay and its rivers (Miyato on the coast, Nobiru in Naruse, and Kadonowaki in Old Kitakami). Map courtesy of OpenStreetMap contributors. From Tolkova (2016)

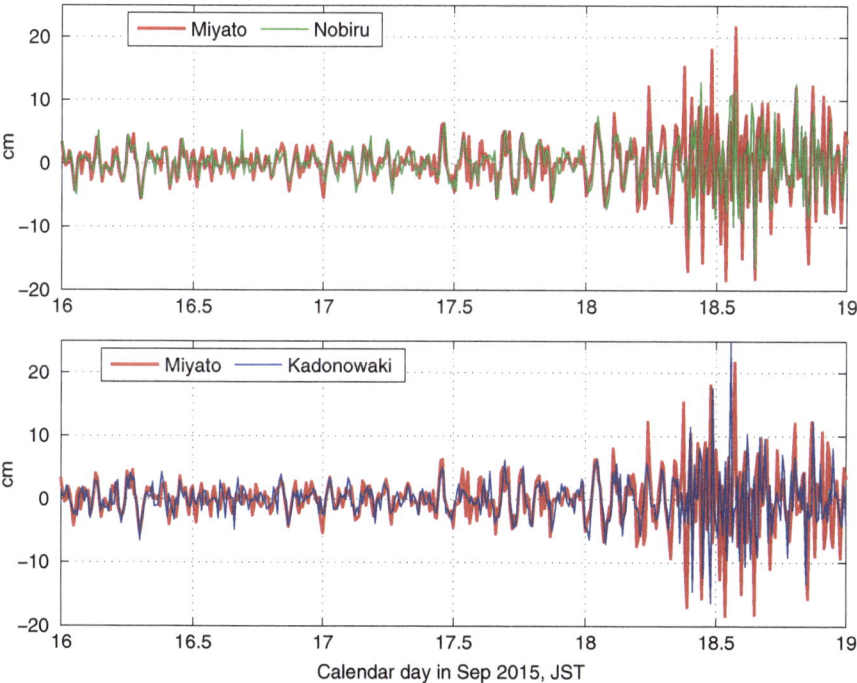

Fig. 6.5 Over-laid segments of de-tided records at Miyato and Nobiru (top), and Miyato and Kadonowaki (bottom). De-tiding residuals prior to the tsunami arrival are the same at all three stations, which suggests that the residuals reflect large-scale normal oscillations in coastal basins. From Tolkova (2016)

A likely cause for the residual background signals in Mabechi and Naka rivers is seiching in coastal waters as well, as detailed in Tolkova (2016). The Mabechi River enters the Hachinohe harbor, and the Naka River enters the Kashima nada.

The tsunami's spectrum is localized between periods from 25 to 40 min. The residual noise is dominated by oscillations at 40–120 min periods. These spectra are well separated from anticipated tidal spectra, which facilitates isolating the tidal component by filtering.

6.4 Correlation Between an Instant Tsunami Amplitude and WLS

Tolkova et al. (2015) suggested that the observed tsunami modulation by tide is, in fact, a correlation between two quantities: an attenuation rate of a tsunami's wave element amplitude with a traveled distance, and a tidally-modified WLS in the element's path. Tolkova (2016) verified this proposition with the records of the Chile 2015 tsunami at 12 stations in five rivers (Fig. 6.2). The rest of this chapter follows the latter study.

Figures 6.6, 6.7, 6.8 display segments of de-tided records at 12 stations in five rivers, as well as the corresponding time-histories of the instant wave amplitudes. For each upstream station, the figures also include WLSs traveled by tsunami/seiche wave elements from the most downstream station in that river. The WLSs are computed with the tide-and-current components of the original records, using travel times listed in Table 6.1. The travel times (in 10 min increments) between the stations are assumed independent of a tidal phase in a river, and of other river conditions. The coarse sampling interval of 10 min readily justifies for this assumption.

As clearly seen in Figs. 6.6, 6.7, 6.8, both tsunami and seiche are modulated according to tidal WLS. The modulation shows up already at the downstream stations after traveling 0.5–1.2 rkm from the mouth, and progresses upriver. In every upstream record, an increase in the WLS traveled by the wave always coincides with the drop in the wave amplitude. Farther upriver, the intruding wave reduces to a few-hour-long trains admitted only at the lowest WLSs (see Ono in Naruse and Kashimadai in Yoshida in Fig. 6.6, or Wabuchi in Old kitakami in Fig. 6.8).

The largest WLSs—and the anticipated lowest levels of rivers' admittances to ocean waves—occurred during a high runoff on September 18–19. The simultaneous occurrence of the tsunami allowed us to probe larger WLSs with a larger signal.

When not affected by weather events, WLS is set by the tide. However, WLS variations do not exactly follow tidal variations. In particular, when the tide is predominantly semi-diurnal, then WLS has two peaks per day as well. The WLS peaks occur shortly before the low water. However, WLS has only one peak per day for both diurnal and mixed tide, which occurs only when the tide transitions to the lower low water. By contrast, WLS remains low when the mixed tide recedes to its

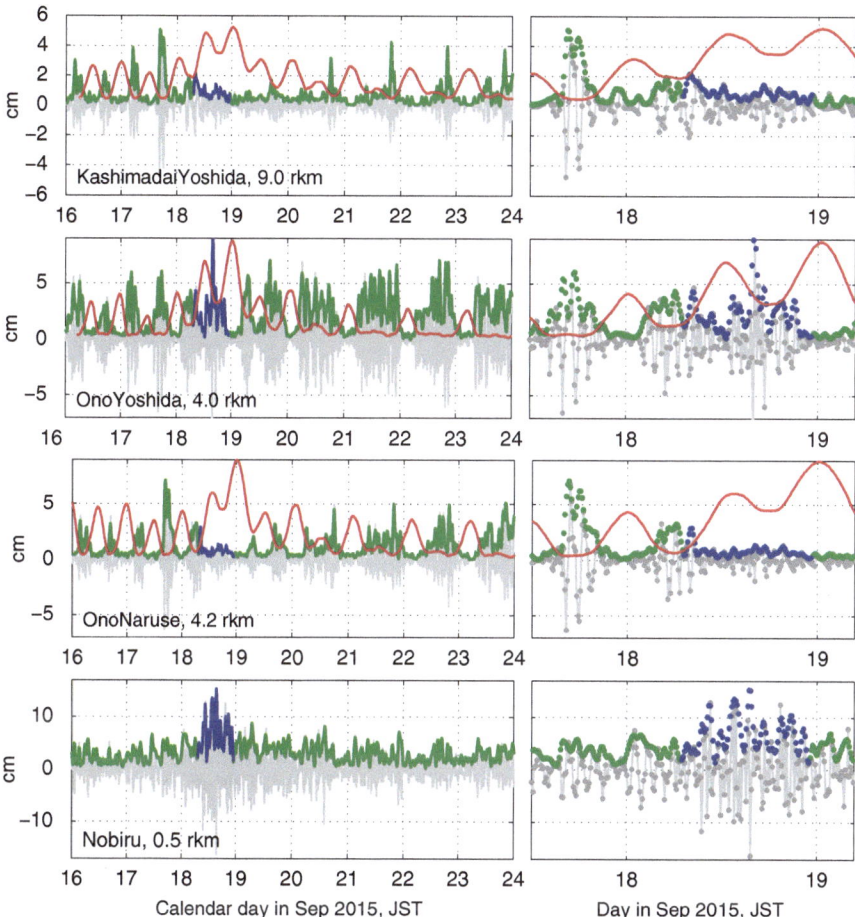

Fig. 6.6 Left: Segments of de-tided records at four stations in Naruse and Yoshida rivers: de-tided measurements (gray), amplitudes of the associated wave elements (green—due to long-wave ocean noise, and blue—for 16 h after the tsunami arrival at the coast), and wave-locked slope (red curve; in cm/rkm, with a factor 0.25) between each upriver station and the station near the river mouth (Nobiru). Right: the same records zoomed-in, with dots representing individual readings or wave elements. The higher the WLS (red), the lower the amplitudes (green/blue). From Tolkova (2016)

second (higher) low level. In our records, the tide evolved from semi-diurnal before the tsunami to the mixed tide after. Consequently, the ocean noise admitted upriver before and after the tsunami event is modulated in different ways, from two intense wave clusters (before) to a single one per day (after), as clearly seen in records at Ono/Naruse or Kashimadai/Yoshida in Fig. 6.6.

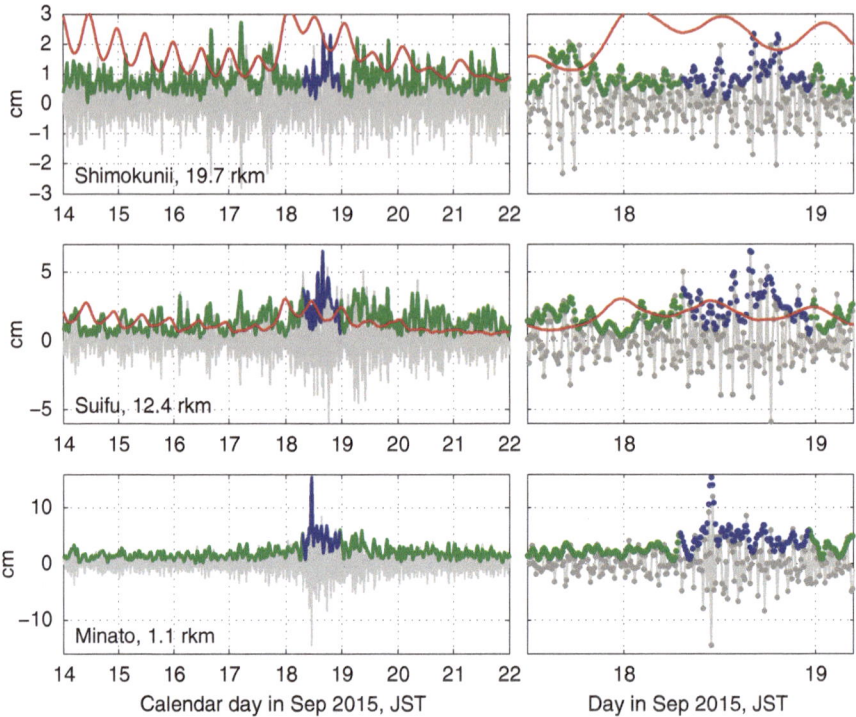

Fig. 6.7 Segments of de-tided records at three stations in Naka River, zoomed-in on the right: de-tided measurements (gray), amplitudes of the associated wave elements (green—due to long-wave ocean noise, and blue—for 16 h after the tsunami arrival at the coast), and wave-locked slope (red curve; in cm/rkm, with a factor 0.5 in the upper panes, and 1.0 in the middle panes) between the respective upriver station and the station near the river mouth (Minato). Individual readings or wave elements are shown with dots in zoomed-in records on the right. The higher the WLS (red), the lower the amplitudes (green/blue). From Tolkova (2016)

6.5   Methodology of Admittance Computations as a Function of WLS

As demonstrated with the field observations, wave elements ascending different surface slopes experience different dissipation. A simple mathematical expression for the above is that an element's amplitude at upriver station B (output) is proportional to its earlier amplitude at station A downstream (input) with an admittance factor κ depending on a traveled WLS β_{AB}:

$$\alpha_B(t) = \kappa(\beta_{AB}(t)) \cdot \alpha_A(t - \tau) \tag{6.3}$$

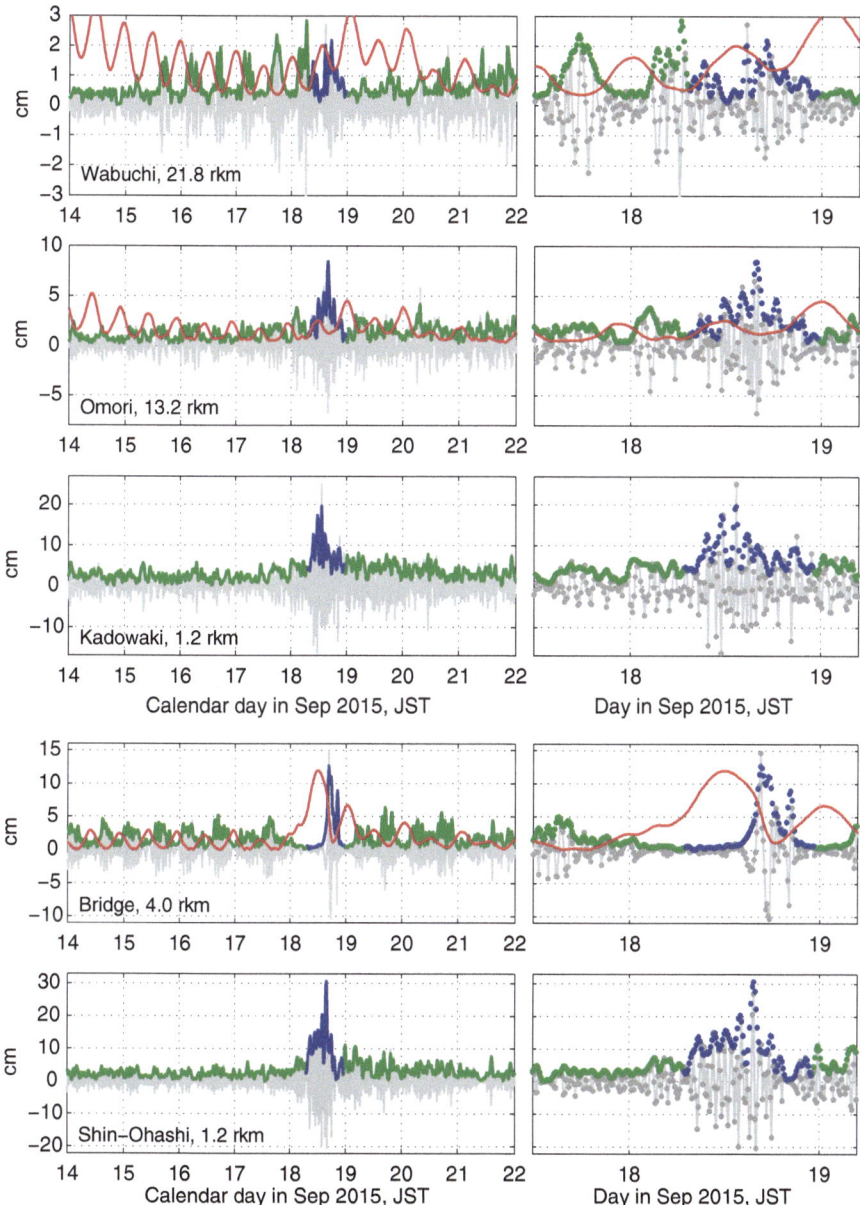

Fig. 6.8 Left: Segments of de-tided records at three stations in Old Kitakami and two stations in Mabechi rivers: de-tided measurements (gray), amplitudes of the associated wave elements (green—due to long-wave ocean noise, and blue—for 16 h after the tsunami arrival at the coast), and wave-locked slope (red curve; in cm/rkm, with a factor 0.5 in Old Kitakami, and a factor 0.25 in Mabecni) between each upriver station and the station near the river mouth (Kadonowaki, Shin-Ohashi). Right: the same records zoomed-in, with dots representing individual readings or wave elements. From Tolkova (2016)

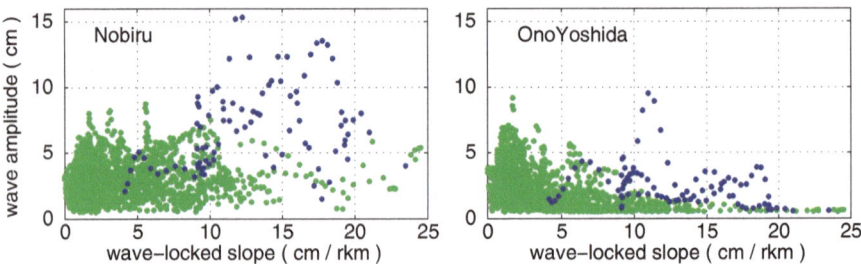

Fig. 6.9 Wave element amplitudes at Nobiru at 0.5 rkm (left) and Ono-Yoshida at 4.04 rkm (right), vs. wave-locked surface slopes traveled by each wave element between the two stations. Blue dots represent wave elements in the 2015/09 tsunami event, green dots represent seiche penetrating into the Yoshida River. From Tolkova (2016)

As implied by this expression, an admittance per unit distance is $\kappa^{1/D}$, where D is a distance between A and B. An admittance factor between two stations a distance x apart, connected by the same WLS β_{AB}, is $\kappa^{x/D}$.

In Fig. 6.9, wave elements in a 2-week long record (excluding those with amplitudes below 0.5 cm) are mapped according to their amplitudes $\alpha_A(t - \tau)$ at Nobiru (left) or $\alpha_B(t)$ at Ono-Yoshida (right), and traveled WLSs β_{AB} between Nobiru and Ono. As expected, at Nobiru, large amplitudes are observed at all WLSs; while at Ono, large amplitudes are shifted toward small WLSs. Assuming that the noise had the same maximal input amplitudes on all days (and therefore at all WLSs) in the record, one could almost see the dependance $\kappa(\beta_{AB})$ in the Ono plot (Fig. 6.9, right) as an envelope curve to the green dots.

Figure 6.9 provides a clear and a seemingly accurate picture of the admittance being a function of traveled WLSs, in spite of that each dot is likely to be randomly displaced from its correct position with respect to both axes. Firstly, there is a de-tiding error of separating tidal and tsunami components. This error translates into inaccuracies in determining both WLS and an instant tsunami amplitude. Tolkova (2016) argues that in this particular case, 1 cm is a reasonable upper estimate for the de-tiding error. A de-tiding error of 1 cm at one station would produce the same error in the tsunami's element amplitude, and 0.2 cm/rkm error in calculated WLS between stations separated by 5 rkm. Secondly, there is an error of measuring the propagation time τ used for tracking a particular wave element from A to B. The same travel time for a given station pair is used for all wave elements, as listed in Table 6.1. In our case, the propagation time error is determined by the 10-min sampling interval. With a more frequent sampling, this error might be defined by the differences in travel times among wave elements propagating atop different flow depths and currents. The travel time error also translates into inaccuracies in determining both WLS and an element's amplitude.

Therefore, calculating the admittance by applying expression (6.3) to an individual wave element is likely to yield poor results, because every term in (6.3) contains a potential error. Instead, Tolkova (2016) suggested to recover function $\kappa(\beta_{AB})$ from an ensemble of wave elements given by their trios (α_A, α_B, β_{AB}).

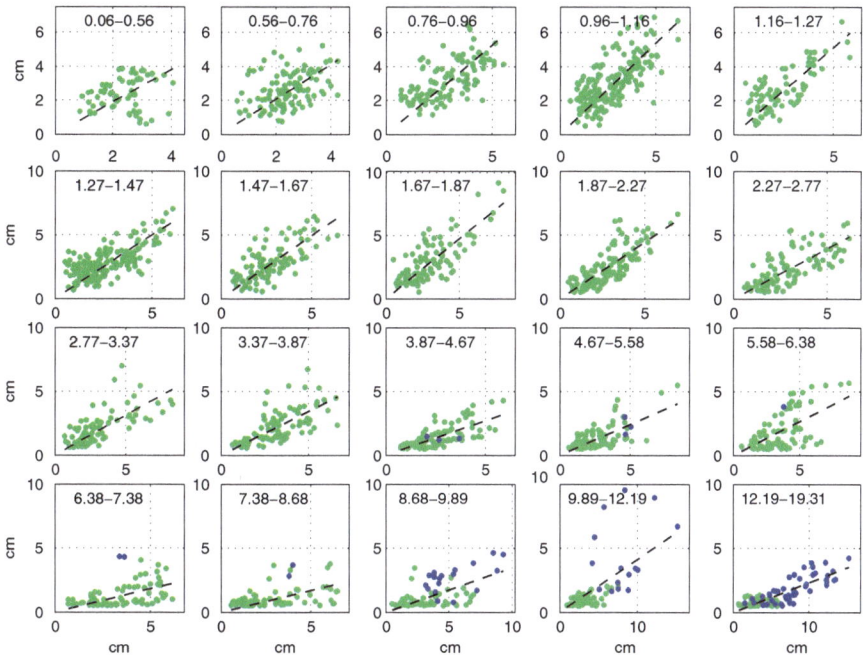

Fig. 6.10 Wave element amplitude at Ono (α_B, *y*-axis) vs its amplitude at Nobiru (α_A, *x*-axis) in the Yoshida river, for all elements traveling in a range of wave-locked slopes indicated in each panel. Blue dots represent wave elements in the 2015/09 tsunami event, green dots represent long-wave ocean noise. Dashed lines in each panel show the best fit with a line $\alpha_B = \kappa \cdot \alpha_A$. From Tolkova (2016)

The procedure was based on sorting wave elements according to traveled WLS, as in Fig. 6.9. Then a range of WLS was divided into uneven intervals, such that each interval contained no fewer than 1/25 of the wave elements in the ensemble (a complete 2-week-long record with a 10-min sampling contains 2016 elements fewer the excluded ones with amplitudes below 0.5 cm). In this manner, wave elements, which traveled between Nobiru and Ono-Yoshida, were sorted into 20 bins corresponding to different WLS intervals. Wave elements, which traveled between other station pairs in other rivers, were sorted into 19–23 bins.

Figure 6.10 displays wave elements ascending in Yoshida, sorted into twenty WLS bins. Each dot plot shows wave elements in a single bin, mapped by their input amplitudes at Nobiru (*x*-axis) and output amplitudes at Ono (*y*-axis). With expected proportionality between the output and the input at a given WLS, the dots are anticipated to fill in a sector, for two reasons. Firstly, WLS in each plot is not constant, but varies within a certain interval. Secondly, the amplitudes contain random errors. So, the admittance factor κ is found by fitting the dots with a line $\alpha_B = \kappa \cdot \alpha_A$, and attributed to the mean WLS in the bin. An admittance obtained in this manner reflects only those wave transformations which took place between Nobiru and Ono. It is not affected by any previous modulation which the wave might have undergone by Nobiru.

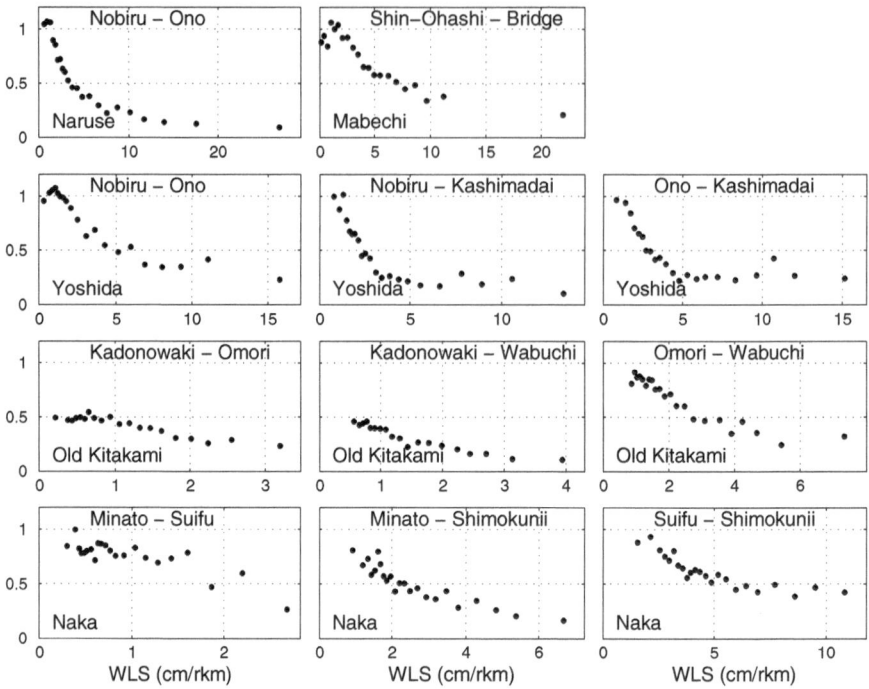

Fig. 6.11 Tsunami admittance factor between two stations as a function of wave-locked slope (black dots). Respective stations and rivers are shown in the plots. From Tolkova (2016)

In this manner, admittance functions $\kappa(\beta)$ have been calculated for every station pair in the five rivers. The resulting admittances at different WLS are shown in Fig. 6.11. Apparently, between every two stations, the admittance co-varies with WLS, decreasing as WLS increases (a few occurrences of an admittance drop at very small WLS will be discussed later). At small WLS, admittance is near 100%, except that between Kadonowaki downstream Old Kitakami and any upriver station (Omori, Wabuchi). There, the admittance does not exceed 50%, while between Omori and Wabuchi, the admittance at low WLS is at 100% again. A likely explanation for this mystery is a presence of a physical obstacle between Kadonowaki and Omori responsible for the loss of nearly a half of the signal—a large island partially barricading the river.

6.6   When Is a Tsunami's Admittance into a Tidal River Greater?

As found in the field data, when tide creates a small WLS in a river, then a co-propagating tsunami (only small tsunamis are considered here) experiences virtually no attenuation, even though the tide might attenuate significantly. For example, the tidal height reduces by a factor of 10 between Nobiru and Kashimadai stations in Yoshida; but the ocean noise remains as high at Kashimadai as it was at Nobiru, when it travels at WLS under 2 cm/rkm. Such a low WLS happens for about 2 h once or twice during the day. Between Nobiru and Ono-Yoshida, during the periods of mixed tide, WLS remains low for as long as a half of the day. At that time, the ocean noise propagates to Ono with no attenuation, while the tide loses about 1/3 of its downstream height. As WLS increases, however, tsunami dissipates at a greater rate. At large WLS, a tsunami can dissipate faster than tide. For instance, from Nobiru to Ono-Naruse, tide reduces in height by a factor of four, while the tsunami amplitude might diminish ten times (with about 10% admittance at a WLS of 27 cm/rkm).

It might be interesting co compare admittance functions of WLS in different rivers, as well as those functions calculated in different segments of the same river. The distance between gauging stations D varies from 2.8 to 20.5 rkm. Hence prior to the comparisons, the admittances are re-calculated to a fixed distance of 10 rkm as $\kappa_{10} = \kappa_{AB}^{10/D}$—for all station pairs, except for Kadonowaki with an upstream station. An island upstream Kadonowaki cuts the intruding wave amplitude in half. This localized impact does not scale with a traveled distance, and it is not present in other channel segments. For inter-river admittance comparisons, we wish to exclude the influence of this island. Without the island, the admittance of a 10-km-long river segment upstream Kadonowaki would be $\kappa_{10} = (2\kappa_{AB})^{10/D}$.

The resulting functions are displayed in Fig. 6.12. Admittances computed for different segments in the same river are plotted in the same axes. Our first observation is that at the same WLS, the larger rivers (Naka and Old Kitakami) provide higher admittances for seiche and tsunami. For instance, after propagating a hypothetical 10 rkm distance at 5 cm/rkm WLS, tsunami wave height would reduce by 50% in Naka, by 75% in Old Kitakami, 85% in Yoshida and Mabechi, and by more than 95% in Naruse. Secondly, we observe a surprisingly good match among the admittance curves computed in different segments in the same river. This latter observation might somewhat contradict the first one, since a river's depth and width are greater in its downstream segment, than between upstream stations. Which river parameter primarily affects its admittance at a given WLS remains an open question. However, Tanaka and co-authors (2014) found that a tsunami's intrusion distance correlates well to the riverbed slope: the milder the slope, the greater the intrusion distance. Admittances found here are also greater in rivers with milder bed slopes (subject to availability of the bed elevation measurements).

6.7 Physics of Admittance Variations with WLS

The above analysis of the field data is silent about the physical reason for why tsunami's attenuation in a river correlates with tidally-modified WLS. We'll try to answer this question by physical reasoning. The most apparent physical mechanism

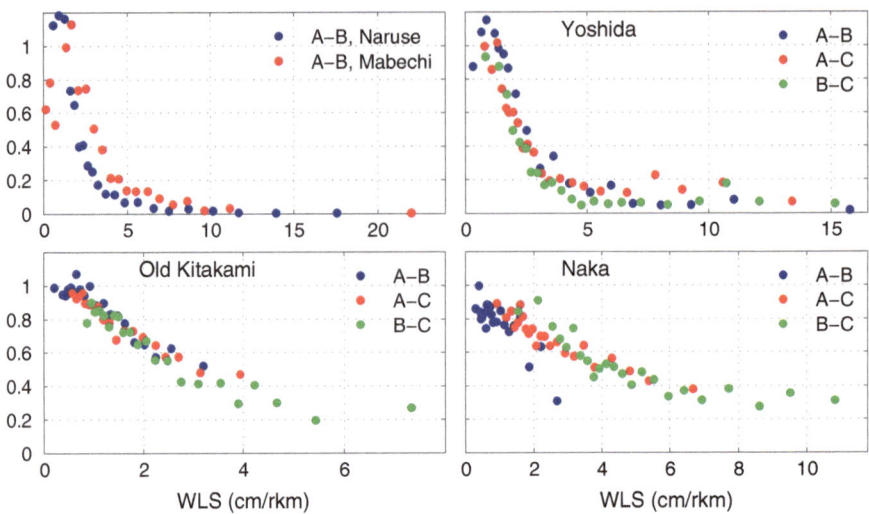

Fig. 6.12 Tsunami admittance factor vs WLS, recalculated to a fixed travel distance of 10 rkm. In rivers with three stations, A refers to the most downstream gauging station, C—to the most upstream; colors refer to computations in segments AB (blue), AC (red), and BC (green). Modified from Tolkova (2016)

for dissipating a wave in a channel is the bed friction. In a river hosting only a steady flow, a stationary surface slope—also known as the hydraulic gradient—creates pressure which balances the bottom friction and supports the flow. In this case, a surface slope is a measure of the bed friction. In Sect. 5.2 of Chap. 5, we found that equations describing a unidirectional wave in a channel with negligible reflection resemble those describing a steady flow, if the flow is mapped by an observer moving with the wave. Should this observer measure the surface elevation or the flow velocity, the measurements in different locations will be taken at different times, as the wave passes those locations. A surface slope measured by this observer comprises the WLS. Nevertheless, the equations governing the flow are alike regardless of whether the flow variables are measured in space simultaneously, or measured along the space-time wave path. Consequently, WLS measures local friction under an intruding wave, likewise a surface slope measures friction under a regular flow. Different from the true hydraulic gradient, however, WLS does not entirely translate into pressure, and therefore does not prevent the wave from dissipating.

Thereby tidally-induced WLS correlates with local friction under a particular tidal phase. For well subcritical flows (with flow velocity V much smaller that the wave propagation speed), WLS is simply f/g, where f is the depth-averaged particle acceleration due to bottom friction, referred to for brevity as the friction term, and g is the gravity acceleration. Local friction experienced by tidal wave is commonly considered proportional to the square of the flow velocity: $f(V) = -\gamma V|V|$, where γ is the proportionality factor characterizing, in particular, the bed roughness. Under the same frictional model, local friction experienced by combined tide and tsunami flows is $f(V + v) = -\gamma(V + v)|V + v|$, where V and v are tidal and tsunami flow components, respectively. For a small-amplitude tsunami, $|v| < |V|$ at most tidal phases in the tidal cycle. Then $sign(V + v) = sign(V)$, and the friction term in the combined flow can be expanded as in Sect. 5.5:

$$f = -\gamma|V + v|(V + v) = -\gamma\frac{|V|}{V}(V + v)^2 = -\gamma|V|V - 2\gamma|V|v + O(v^2) \quad (6.4)$$

The first term in the right part of (6.4) represents friction acting upon the tidal component. The next term varies on the tsunami time scale, and represents friction acting upon the tsunami component, as elaborated in more detail in Sect. 5.5. For $|v| \ll |V|$, the last term is negligible. The first term can be estimated with the tidal WLS χ_s as

$$f(V) = -\gamma|V| \cdot V \approx g\chi_s. \quad (6.5)$$

Then the second term can be expressed through the tidal WLS as well:

$$- 2\gamma|V|v = -2\sqrt{g\gamma|\chi_s|}v. \quad (6.6)$$

Hence friction upon the tsunami component is proportional to the tsunami's flow velocity, with the proportionality coefficient varying with WLS set by the co-propagating tide. That explains why the tsunami's attenuation in a river correlates with tidally-modified WLS. It seems, that a typical pattern of WLS variation over a tidal cycle exists in tidal rivers: WLS is at its lowest near the high tide, and at its highest near the low tide. In other words, the surface elevation at low tide tends to take a much steeper upriver path, than that under high tide, as demonstrated analytically in Sect. 5.4 of the previous chapter. WLS is also asymmetric on receding and rising tide, being greater when the tide recedes. This WLS behavior must be common in all rivers which exhibit the described pattern of the tsunami's modulation by tidal phase.

Lastly, we explain an unexpected admittance drop at the very small WLS, observed in some cases in Figs. 6.11 and 6.12. An average WLS (6.2) between stations A and B relates to the local WLS as

$$\beta_{AB} \cdot (x_B - x_A) = \int_{x_A}^{x_B} \beta dx, \quad (6.7)$$

where β is measured along the wave path (that is, by an observer moving with the wave). Should β change its sign as the wave propagates from A to B, an average WLS would underestimate an absolute local WLSs between the stations. For instance, WLS changes its sign over a path of a tidal crest entering with an inflow at A, but meeting an outflow at B (due to the riverine flow contribution). Then frictional losses corresponding to the higher local WLSs will be mapped to a very small average WLS, creating an artificial drop of the measurement-derived admittance.

References

Kayane, K., Min, R., Tanaka, H., & Tinh, N. X. (2011). Influence of river mouth topography and tidal variation on tsunami propagation into rivers. *Journal of Japan Society of Civil Engineers, Series B2 (Coastal Engineering)*, *B2–67*(1), I_246-I_250 [in Japanese].

Tanaka, H., Kayane, K., Adityawan, M. B., Roh, M., & Farid, M. (2014). Study on the relation of river morphology and tsunami propagation in rivers. *Ocean Dynamics*, *64*(9), 1319–1332. https://doi.org/10.1007/s10236-014-0749-y.

Tolkova, E. (2013). Tide-tsunami interaction in Columbia river, as implied by historical data and numerical simulations. *Pure and Applied Geophysics*, *170*(6), 1115–1126. https://doi.org/10.1007/s00024-012-0518-0.

Tolkova, E. (2016). Tsunami penetration in tidal rivers, with observations of the Chile 2015 tsunami in rivers in Japan. *Pure and Applied Geophysics*, *173*(2), 389–409. https://doi.org/10.1007/s00024-015-1229-0.

Tolkova, E., Tanaka, H., & Roh, M. (2015). Tsunami observations in rivers from a perspective of tsunami interaction with tide and riverine flow. *Pure and Applied Geophysics*, *172*(3–4), 953–968. https://doi.org/10.1007/s00024-014-1017-2.

Wilson, B. W., & Torum, A. (1972). Effects of the tsunamis: An engineering study. In *The Great Alaska earthquake of 1964: Oceanography and coastal engineering* (Committee on the Alaska Earthquake, National Research Council) (pp. 361–526). Washington, DC, USA: National Academy of Sciences.

Yeh, H., Tolkova, E., Jay, D., Talke, S., & Fritz, H. (2012). Tsunami hydrodynamics in the Columbia river. *Journal of Disaster Research*, *7*(5), 604–608.

Chapter 7
Tsunami Bores

Highlights A wave with a vertical front. Shock equations. Computing flow speeds with flow depths before and after the shock. Giving up momentum correction coefficients. Shock conditions in a channel with vertical banks. Shock conditions in a channel with sloping shores. How does an ordinary wave become a bore? Turbulence or ripples? Shock reflection from a barrage: observations and analytics. How good is this theory? Bore train passage up and down the Kitakami River in the 2011 Tohoku tsunami event: what the theory tells, and the records imply. Summary of the book's main points.

7.1 Shock Equations

Large tsunamis have been observed in rivers as shock waves, or bores—a wave with a nearly vertical front separating two different water levels. In spite of irregular shapes of natural river beds, the observed shock fronts tend to maintain uniform elevation across the river.

The SWE indeed allow a solution containing a moving vertical front. In fact, in a frictionless flat channel, a sequence of two uniform flows connected through a discontinuity is the only solution to the SWE which can propagate at a constant celerity, without change in shape. The flow parameters on both sides of the discontinuity are constrained by so called shock equations/conditions. The shock conditions are obtained either by applying mass and momentum balance to a hypothetical fluid block encompassing the shock front,[1] or directly from the SWE which stem from the same physical concepts (Stoker 1957; Henderson 1966; Chanson 2012).

With natural rivers in mind, the shock conditions will be obtained here from the cross-river integrated SWE (2.5) and (2.10) discussed in the first section of the First Analytical Chapter. These equations, reproduced below, had been derived

---

[1]The fluid block, however, moves with respect to the shock, so holding the shock front inside the block might appear challenging, if the block's length dx tends to zero (as it does, eventually).

© The Author(s) 2018

E. Tolkova, *Tsunami Propagation in Tidal Rivers*, SpringerBriefs
in Earth Sciences, https://doi.org/10.1007/978-3-319-73287-9_7

for straight channels under the shallow-water approximation. The only assumption made during the derivation is that the surface elevation $\eta(x, t)$ is constant across the river. The momentum equation (2.10) is expressed in terms of the original—as different from cross-sectionally averaged—bed elevation $\tilde{d}(x, y)$, velocity $\tilde{u}(t, x, y)$, and flow depth $\tilde{h}(t, x, y) = \tilde{d} + \eta$, mapped by along-river x and cross-river y coordinates:

$$A_t + (Au)_x = 0 \tag{7.1}$$

$$\int_{y_1}^{y_2} \left((\tilde{u}\tilde{h})_t + (\tilde{u}^2\tilde{h})_x + (g\tilde{h}^2/2)_x \right) dy = g \int_{y_1}^{y_2} \tilde{h}\tilde{d}_x dy + Af \tag{7.2}$$

where notations are the same as in Chaps. 2 and 5, bars over cross-river averaged values are omitted, original (before averaging with respect to y) values are indicated with the 'tilde' sign. The flow area is defined as

$$A = \int_{y_1}^{y_2} \tilde{h} \cdot dy = b \cdot h \tag{7.3}$$

where $y_1(t, x)$ and $y_2(t, x)$ are the left and right shoreline coordinates, $b(t, x) = y_2 - y_1$ is the river breadth.

We are looking for a hypothetical solution to the SWE with a sharp front, which keeps its shape while propagating at a constant celerity[2] c. Then in the jump's vicinity, all flow parameters depend on x and t as functions of $(x - ct)$. Substituting $\partial/\partial t = -c \cdot \partial/\partial x$ in Eqs. (7.1)–(7.2) yields:

$$(A(u - c))_x = 0 \tag{7.4}$$

$$\frac{\partial}{\partial x} \int_{y_1}^{y_2} \left(\tilde{u}(\tilde{u} - c)\tilde{h} + g\tilde{h}^2/2 \right) dy = g \int_{y_1}^{y_2} \tilde{h}\tilde{d}_x dy + Af \tag{7.5}$$

In terms of the flow velocity $\tilde{v} = \tilde{u} - c$ in a system of reference moving with the jump, and its respective cross-river average $v = u - c$, (7.4) and (7.5) become:

$$(Av)_x = 0 \tag{7.6}$$

$$\frac{\partial}{\partial x} \int_{y_1}^{y_2} \left((c + \tilde{v})\tilde{v}\tilde{h} + g\tilde{h}^2/2 \right) dy = g \int_{y_1}^{y_2} \tilde{h}\tilde{d}_x dy + Af \tag{7.7}$$

The last equation, written in terms of cross-river averages, reads:

$$\left(cvA + mAv^2 + g\hat{h}A/2 \right)_x = A(gd_x + f) \tag{7.8}$$

---

[2]By this assumption, undular bores are excluded from consideration, since the undulations propagate at a different celerity than that of the bore front. Undular bores are treated under a different mathematical framework, such as that in Tsuji et al. (1991) and Yasuda (2010).

where the bed slope d_x was assumed constant across the river, for simplicity; m is the momentum correction coefficient *in the moving system of reference*, and

$$\hat{h} = \frac{1}{A} \int_{y_1}^{y_2} \tilde{h}^2 dy. \tag{7.9}$$

Since the wave celerity c is considered constant, then the first term in (7.8) is zero according to (7.6). Finally, integrating (7.6) and (7.8) across the jump between points x_0 and x_1 yields:

$$A_0 v_0 = A_1 v_1 = M \tag{7.10}$$

$$m_1 A_1 v_1^2 - m_0 A_0 v_0^2 + (g/2)(\hat{h}_1 A_1 - \hat{h}_0 A_0) = \int_{x_0}^{x_1} A(g d_x + f) dx \tag{7.11}$$

where a constant M, uniform along the river, represents a volume flux in a system of reference moving with the wave; indexes refer to flow parameters at points x_0 and x_1, respectively.

Should the bottom elevation vary continuously, the right part of (7.11) tends to zero, if both x_0 and x_1 approach the shock front. Therefore, the flow parameters immediately before and after the jump must satisfy:

$$A_0 v_0 = A_1 v_1 = M \tag{7.12}$$

$$m_1 A_1 v_1^2 - m_0 A_0 v_0^2 = (g/2)(\hat{h}_0 A_0 - \hat{h}_1 A_1) \tag{7.13}$$

Relations (7.12)–(7.13) are known as shock conditions. Should the flow be cross-river uniform ($m_0 = m_1 = 1$), the two relations connect five parameters: the shock celerity c, the flow velocities u_0 and u_1, and the flow areas in front and behind the shock. Should any three of these parameters be obtained from measurements, the other two can be calculated using the shock conditions. Note, that "another average" flow depth $\hat{h}$ (which is different from a common sense average depth $h = A/b$, though $\hat{h}$ and h coincide in a channel with a rectangular cross-section) is not independent of the flow area, but both are determined by the surface elevation and the cross-river bed profile.

7.1.1 Giving Up Momentum Correction Coefficients...

Along-river water level measurements of an ascending bore permit to evaluate the flow depth and areas before and after the shock. The shock conditions can then be used to calculate a volume flux M relative to the shock. To facilitate those computations, the left part of Eq. (7.13) is modified using (7.12) as:

$$m_1 A_1 v_1^2 - m_0 A_0 v_0^2 = m_1 \cdot A_0 v_0 \cdot v_1 - m_0 \cdot A_1 v_1 \cdot v_0 =$$

$$= v_0 v_1 (m_1 A_0 - m_0 A_1) \tag{7.14}$$

Substituting (7.14) into (7.13), multiplying the resulting equation by A_1A_0, and using (7.12) again, yields an equation for the discharge rate M across the shock:

$$M^2 = (g/2)A_0A_1(\hat{h}_1A_1 - \hat{h}_0A_0)/(m_0A_1 - m_1A_0), \tag{7.15}$$

with a sign of M being opposite to a sign of the bore's propagation speed—that is, $M < 0$ for an upriver-going bore ($c > 0$), and $M > 0$ for a bore traveling downstream ($c < 0$). The flow height and area are always greater behind the shock than in front of it, that is, $\hat{h}_1 > \hat{h}_0$ and $A_1 > A_0$. Then the average velocities u_0 and u_1, and the inflow rates Q_0 and Q_1 relative to the river banks, are found as

$$u_j = c + M/A_j, \quad Q_j = cA_j + M, \quad j = 0, 1. \tag{7.16}$$

In particular, if a bore is advancing into still water, then $u_0 = 0$, and consequently, the bore's celerity is $c = -M/A_0$. Also in this case, $m_0 = 1$, while m_1 depends on the bed features, possibly causing complex flow patterns. If $m_1 = A_1/A_0$ (which won't be far from unity for a shock with a small height), then (7.15) yields an infinite M and an infinite bore's celerity c.

The paradox originates with manipulating cross-sectional velocity distribution (coefficient m_1) independently of the cross-flow shape, which stems from the primarily assumption behind the cross-river averaging—that the surface elevation is horizontal across the river. The truth is that the cross-river averaged theory applies best to cross-river uniform flows. How accurate can then the theory be when it is used to describe a real tsunami in a real river? Tolkova and Tanaka (2016) tried to answer this question by exercising the shock equations with water level measurements in a train of tsunami bores, which traveled up and down the Kitakami River in the 2011 Tohoku event. They avoided over-sensitivity to values of m_0 and m_1 by imposing $m_0 = m_1 = m \geq 1$, thus considering m as a mere scaling-down factor for the discharge rate M. Below, we elaborate upon the work of Tolkova and Tanaka (2016), tackling the same question: whether and how the cross-river averaged shallow-water approximation can be adapted for interpreting tsunami flows in the real-world rivers?

7.1.2   Shock Conditions in a Channel with Vertical Banks

In a channel with vertical banks, the flow width b is the same on both sides of the shock. Then

$$\hat{h}_1A_1 - \hat{h}_0A_0 = \int (\tilde{h}_1^2 - \tilde{h}_0^2)dy = (\eta_1 - \eta_0)\int (\tilde{h}_1 + \tilde{h}_0)dy =$$
$$= (\eta_1 - \eta_0)(A_1 + A_0), \tag{7.17}$$

since η is constant across the river. Furthermore,

$$A_1 - A_0 = \int (\tilde{h}_1 - \tilde{h}_0)dy = (\eta_1 - \eta_0) \cdot b. \tag{7.18}$$

Then with $m_0 = m_1 = m$, (7.15) simplifies to

$$M^2 = gA_0A_1 (A_0 + A_1) / (2mb) \tag{7.19}$$

Using (7.14), the second of the shock conditions (7.13) simplifies as well, and becomes:

$$v_0v_1 = g(A_0 + A_1)/(2mb). \tag{7.20}$$

In particular, in a channel with a rectangular cross-section hosting a uniform flow, $m = 1$, $\hat{h} = \tilde{h} = h$, and $A_{0,1} = bh_{0,1}$. Then the shock conditions simplify to their most known form:

$$h_0v_0 = h_1v_1, \quad v_0v_1 = g(h_0 + h_1)/2, \tag{7.21}$$

expressing preservation of mass and momentum in a single vertical plane.

7.1.3 Shock Conditions in a Channel with Sloping Shores

In a channel with linearly sloping shores above the water surface (Fig. 7.1), the left side of (7.13) can be evaluated as:

$$\hat{h}_1A_1 - \hat{h}_0A_0 = \int_{y_1}^{y_2} (\tilde{h}_1 - \tilde{h}_0)(\tilde{h}_1 + \tilde{h}_0)dy +$$

$$+ \int_{y_1-l_1}^{y_1} \tilde{h}_1^2 dy + \int_{y_2}^{y_2+l_2} \tilde{h}_1^2 dy = I_1 + I_2 + I_3,$$

$$I_1 = \delta\eta(A_0 + A_1 - a_1 - a_2),$$

$$I_2 = \int_0^{l_1} (\delta\eta \cdot y/l_1)^2 \, dy = \delta\eta^2 l_1/3 = (2/3)\delta\eta \cdot a_1,$$

$$I_3 = (2/3)\delta\eta \cdot a_2,$$

hence

$$\hat{h}_1A_1 - \hat{h}_0A_0 = \delta\eta (A_0 + A_1 - (a_1 + a_2)/3) \tag{7.22}$$

where l_1 and l_2 are inland inundation distances at the left and right shores, respectively; a_1 and a_2 are the flow areas above the left and right shores inundated by the bore (Fig. 7.1); $\delta\eta = \tilde{h}_1 - \tilde{h}_0 = \eta_1 - \eta_0$.

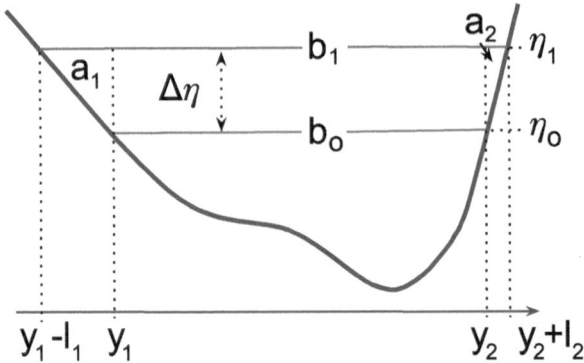

Fig. 7.1 Cross-river dimensions in a channel with sloping shores

Assuming $m_0 = m_1$, the right side of (7.13) can be elaborated as

$$A_1 v_1^2 - A_0 v_0^2 = v_0 v_1 (A_0 - A_1). \tag{7.23}$$

Next geometrically apparent relations between the flow areas hold:

$$A_1 - A_0 = \delta\eta \cdot b, \tag{7.24}$$

where $b = (b_0 + b_1)/2$ is an average width of flows before and after the shock; and

$$a_1 + a_2 = \delta\eta \cdot (b_1 - b_0)/2 = (A_1 - A_0)\Delta b/(2b), \quad \Delta b = b_1 - b_0. \tag{7.25}$$

Using (7.25) with (7.22), and (7.24) with (7.23), condition (7.13) can be formulated as:

$$v_0 v_1 = \frac{g}{2mb}(A_0 + A_1)(1 - \epsilon), \quad \epsilon = \frac{A_1 - A_0}{A_1 + A_0} \cdot \frac{\Delta b}{6b} \tag{7.26}$$

Factor $\Delta b/(6b)$ never exceeds 1/3. If the flow width behind the bore doubles, $\Delta b/(6b)$ is only 1/9. Given that, typically, $A_1 - A_0 \ll A_1 + A_0$ as well, then $\epsilon \ll 1$. The flow width difference across the shock contributes a relatively small correction to the shock equation (7.26), which is very close to an analogous equation (7.20) for a river with a fixed width. It is essential, however, to characterize the flow by a width b equal to an average width before and after the shock.

7.1.4   Whose Flow Is Faster?

Let h_1 and u be a flow depth and a current in a segment of an ordinary wave propagating into a still frictionless channel with a flat bottom and a depth h_0 in the undisturbed state. Then h_1 and u satisfy a known relation based on the theory of characteristics[3] (Stoker 1957):

$$u = 2\sqrt{gh_1} - 2\sqrt{gh_0} = c_0 \cdot 2(\sqrt{\epsilon} - 1), \quad c_0 = \sqrt{gh_0}, \quad \epsilon = h_1/h_0. \qquad (7.27)$$

Next, let h_1 and u_1 be a flow depth and a current in a bore advancing into the same channel. After solving the shock equation (7.21) for v_0 and v_1 ($v < 0$ for a positively directed bore), we obtain

$$v_1 = -c_0 \cdot \sqrt{(1+\epsilon)/(2\epsilon)}, \quad v_0 = v_1 \cdot \epsilon$$

$$u_1 = v_1 - v_0 = c_0 \cdot (\epsilon - 1)\sqrt{(1+\epsilon)/(2\epsilon)}, \quad \epsilon \geq 1. \qquad (7.28)$$

Figure 7.2 shows flow velocity behind a bore and that under an ordinary wave, as a function of flow depth relative to the depth of still water h_1/h_0. Flow speed in a bore slightly exceeds that in a wave of the same height. For instance, the same flow with a $0.866c_0$ speed exists behind a shock with a height equal to the basin depth ($h_1 = 2h_0$), and under a wave crest $1.05h_0$ high ($h_1 = 2.05h_0$).

Fig. 7.2 Relative flow speed u_1/c_0 under a bore (red), and relative flow speed u/c_0 under an ordinary wave (green), as a function of relative flow depth h_1/h_0

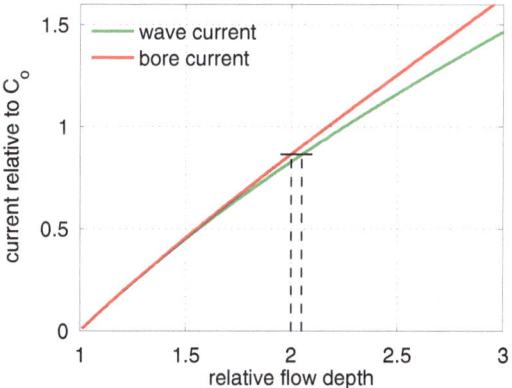

---

[3]Consider a left-going characteristic which connects a quiescent state with parameters h_0 and $u_0 = 0$, not yet reached by the wave, and a state of flow with parameters h_1 and u. In a flat, frictionless basin, the left-going (as well as the right-going) Riemann invariant ζ^l (5.6) remains constant along its characteristic. Equating the values of ζ^l at the two states yields (7.27).

7.2   How Does an Ordinary Wave Become a Bore?

There is little, if any, *evidence* of how a tsunami transforms into a bore. From observations of much shorter wind waves breaking near the beach, shock waves are believed to originate with non-linear wave steepening, followed by the wave breaking. Since a wave celerity is greater in a greater depth, wave crests are catching up with the wave troughs ahead. At the same time, wave troughs fall behind the preceding wave crests. A wave segment behind a trough, where the wave elevation is on the rise, compacts. A wave segment behind a crest, where the wave elevation is falling, expands. This transformation is seen in Fig. 7.3, which shows simulated time histories of a sine wave with a 30 min period propagating into a still frictionless 5-m-deep channel. The time histories are displayed at 1, 5, and 10 km marks into the channel. The earliest vertical front develops behind the trough, where the flow depth is smaller, and therefore relative differences in the propagation speed for different wave elements are greater. Eventually, the numerical wave assumes the saw-tooth shape. The saw-tooth profile is common to numerical models solving the SWE, while its further evolution (not shown), including decomposition into undulations, are specific to the model's numerical scheme, in particular, its dispersive and diffusive properties. In the vicinity of the vertical fronts, neither the numerical solution represents the SWE solution anymore, nor the SWE solution represents the physical world. A real wave cannot develop a finite vertical front from propagating over a flat bottom. Instead, it curls and breaks. A vertical front forms following the breaking (Stoker 1957).

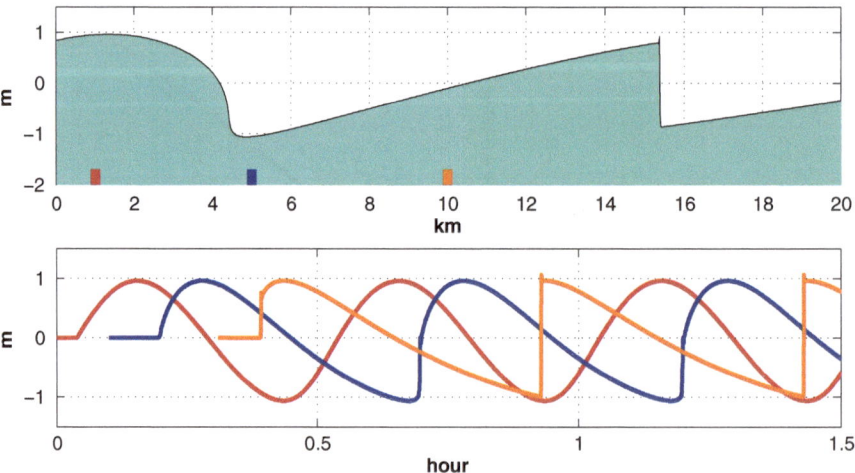

Fig. 7.3 Numerical simulation of a sine wave propagating into a still frictionless 5-m-deep channel. Top: a sample surface profile. Colored bars indicate positions of three virtual gages at 1.0 (red), 5.0 (blue), and 10.0 (orange) km. Bottom: time histories at three gages, color-coded according to respective gages

According to the characteristics form of the SWE, its Riemann invariants—and therefore the flow depth h and velocity u—remain constant along characteristics of a wave propagating over a frictionless, flat bed. Thus consecutive segments of the wave, referred to in Chap. 6 as wave elements, propagate at a constant elevation each, with a constant flow speed in each element. In particular, a crest and a trough propagate at constant elevations and carry constant currents, as well. Thereby, as seen in Fig. 7.3, the wave transforms its shape, but maintains its height.

Once a wave starts to curl, it cannot be described by the SWE in the breaking segment anymore, but the SWE solutions should still be valid close by. As suggested by Stoker (1957), wave breaking "may not seriously affect the motion of water behind it". A newly emerged front borders the flow which existed behind the breaking segment. This flow, therefore, conforms to the SWE. Let h_1 and u_1 be the flow parameters behind the newly emerged bore. It is reasonable to expect that the breaking does not affect the flow speed of the following water mass. This speed u_1 is approximately the speed at the crest before breaking. Then to satisfy the shock conditions, the flow depth h_1 should be slightly below the flow depth at the crest—but only slightly (see Fig. 7.2 in the previous section). Hence under the above transformation mechanism, and with an exception of monstrously large bores, it is suffice to expect that a wave transforms into a bore of the same height. Depth-averaged numerical models appear to successfully reproduce the main course of this transformation, even though they "skip" breaking. They also may not be an adequate tool for predicting further evolution of the developed shock fronts.

Uneven bathymetry (Fig. 7.4) can greatly speed-up the process of transforming an ordinary wave into a bore. Figure 7.5 shows snapshots and time histories of a sine wave with a 30 min period propagating from a 4.5-km deep ocean, over an 80-km wide continental slope, and into a river 5-m-deep at the mouth, carrying 5 m^2/s discharge per 1 m width. Wave elevation in time histories is counted from local water levels before the wave intrusion. Offshore time histories display a great deal of wave amplification due to shoaling, but show no significant deviations from the

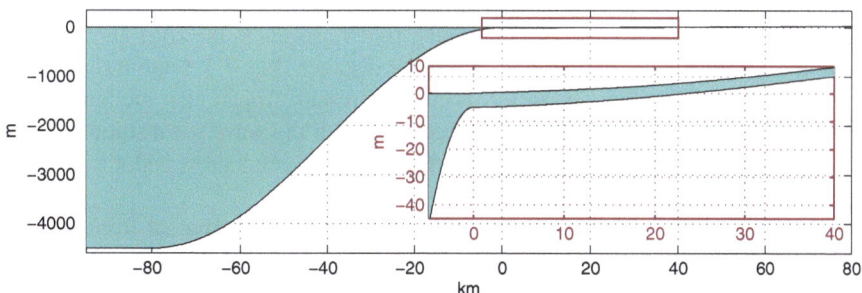

Fig. 7.4 Domain for numerical simulations composed of a river connected to an ocean at $x = 0$: the whole domain view, and a zoom around the river's mouth (inset). River's bed elevation varies as $d(x) = 5 - (x/8000)^{1.5}$; freshwater current at the mouth is 1 m/s; bottom roughness $n = 0.03$ $s \cdot m^{-1/3}$ in Manning's formulation

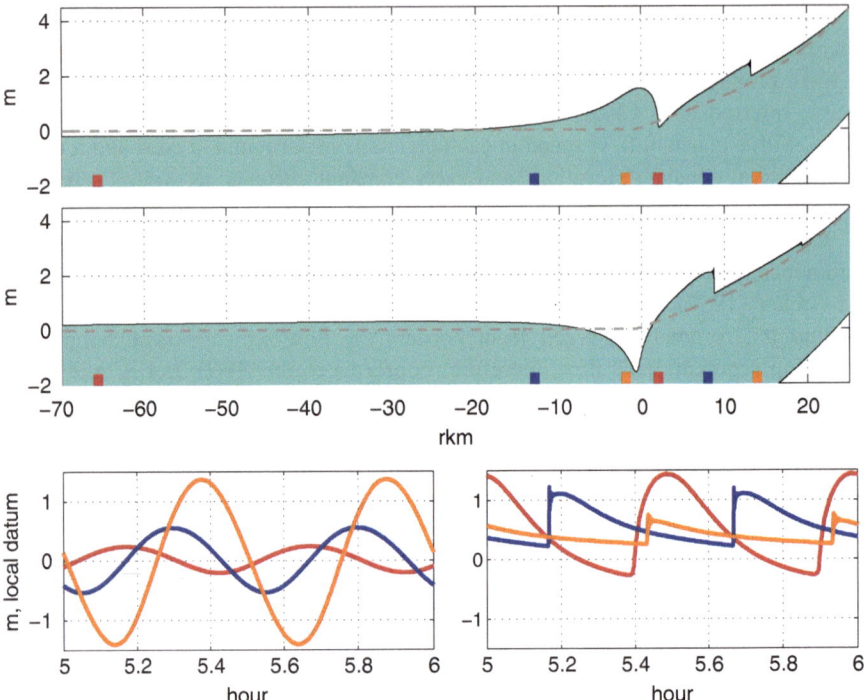

Fig. 7.5 Numerical simulation of a sine wave propagating from an ocean into a river, in a basin shown in Fig. 7.4. Top two panes show surface profiles at the moments when a crest or a trough are approaching the river's mouth. Colored bars indicate positions of three virtual gages in the ocean at 65.7 (red), 13.0 (blue), and 1.9 (orange) km offshore, and three gages in the river at 2.0 (red), 8.0 (blue), and 14.0 (orange) rkm upstream. Bottom left: time histories at three offshore gages, color-coded according to respective gages. Bottom right: time histories at three in-river gages, color-coded analogously. Wave elevation in time histories is counted from local water levels before the wave intrusion, indicated by a dashed gray line in the upper plots

sine law (Fig. 7.5, bottom left). Differently, time histories in the river show a train of bores diminishing in height with an upriver distance (Fig. 7.5, bottom right). The wave set-up discussed in Chaps. 3–4 is clearly present. Apparently, the bulk of the wave's transformation into a bore has occurred with the wave's transition into the river, in a place where the flow depth differences between a crest and a trough are greatly amplified by the bed shape.

Amateur videos[4] show how an undular bore develops a turbulent front without any wave breaking. Instead, the turbulence starts near the banks where the river's depth is most non-uniform, and sharply spreads across the river.

---

[4]https://www.youtube.com/watch?v=8iG-39KvNvk.

7.3 Turbulence or Ripples?

As detailed in textbooks, mechanical energy must be dissipated at the shock (Stoker 1957). Therefore, a shock must be dressed in turbulence or emit ripples, both are considered to be the mechanisms of energy dissipation at the shock, either by transforming it into heat, or by radiating away. The first mechanism (turbulence) can dissipate much more energy, than ripples can carry away. The rate of the energy loss is proportional to a cube of the difference in the flow depths across the shock, and therefore sharply increases with the shock height. Hence undular bores are expected to have a small height, while tall bores are expected to have turbulent fronts. Chanson (2010) presents a laboratory study of flow characteristics in bores propagating in a rectangular channel. As found in this study, undular bores form when a ratio of the flow depth behind the shock to that in front of the shock is below 1.7–2.1. Field observations confirm the tendency of undular bores be smaller, but often display turbulent fronts even on relatively low bores. Something in the nature favors turbulence around the bores. Tolkova and Tanaka (2016) suggest that this something is non-uniformity of natural channels in the cross-channel direction. In a channel with an irregular cross-section, the flow momentum is not preserved in each vertical plane. Therefore, to prevent the bore from disintegrating, the momentum must be constantly redistributed across the river—another task efficiently performed by the turbulence. It might be the turbulence which provides for the bore's stability as it passes trough the river's bends, shoals, and rapids.

7.4 Shock Reflection from a Barrage: Observations and Analytics

As described in the first chapter, a train of shock waves of the 2011 Great Tohoku tsunami propagated for 17.2 rkm up the Kitakami River until it hit a weir, where the bulk of the wave was reflected back. The tsunami caused extensive inundation for about 6 rkm from the mouth. Further upstream, as seen in Fig. 7.6, the 5-m high

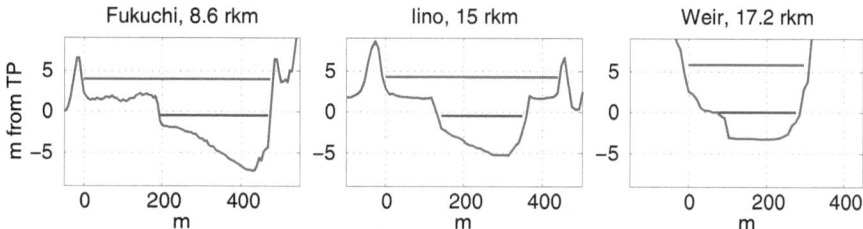

Fig. 7.6 Cross-river profiles at the stations at Fukuchi, Iino, and the weir: gray—bed profile across the river; blue—the water level right before the tsunami arrival, red—the maximal observed level. The illustration uses the Kitakami River bathymetry developed in the National Institute for Land and Infrastructure Management, Japan. From Tolkova and Tanaka (2016)

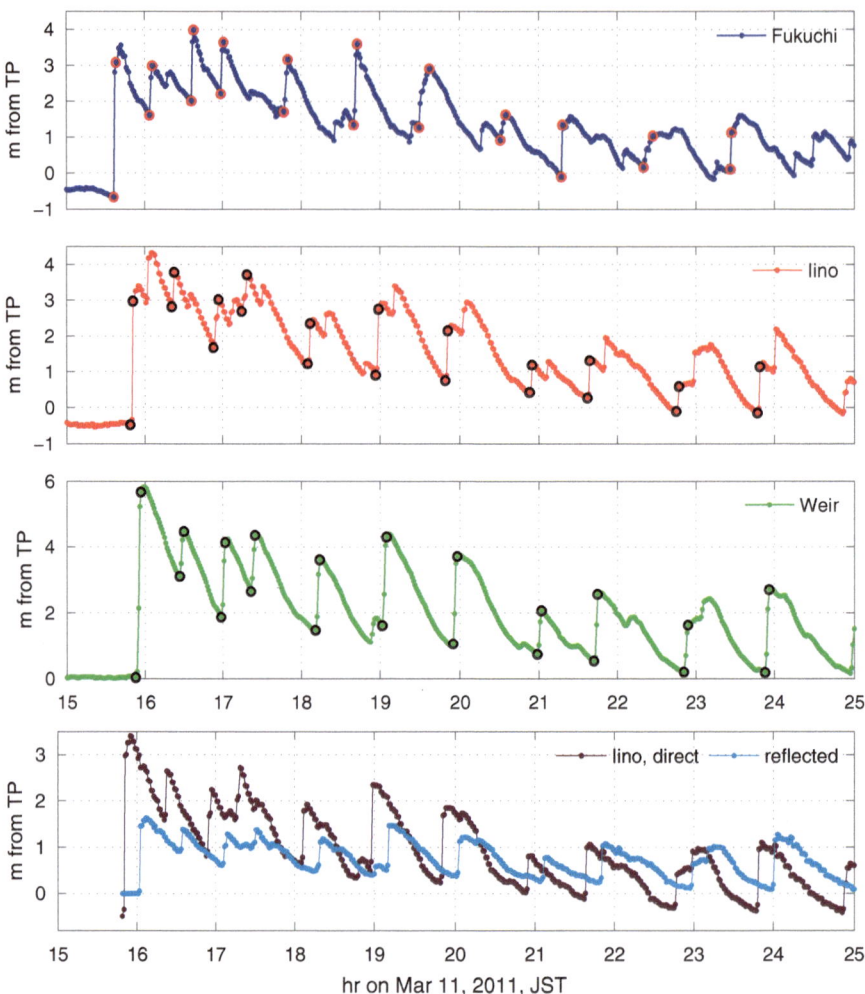

Fig. 7.7 Top three plots: zoomed-in records showing the first eleven tsunami waves at the locations. Circles (red or black) mark readings at the foot and on the top of the first eleven upriver-going shocks. Bottom: Iino record decomposed into direct and reflected wave-trains. From Tolkova and Tanaka (2016)

dikes were able to hold the tsunami within the river's flood plain. The tsunami passage up and down Kitakami was recorded at three locations: at Fukuchi at 8.6 rkm, Iino at 14.9 rkm, and by the weir. The records, shown in Fig. 7.7, permit identifying and tracking the individual bores. Fukuchi record contains only upriver going shocks. Iino record clearly shows direct and reflected bores, which had dissipated before they could reach Fukuchi. Record at the weir shows reflected bores superposed atop the direct ones. Nevertheless, under the cross-river averaged shallow-water approximation, the latter record allows to determine the heights of

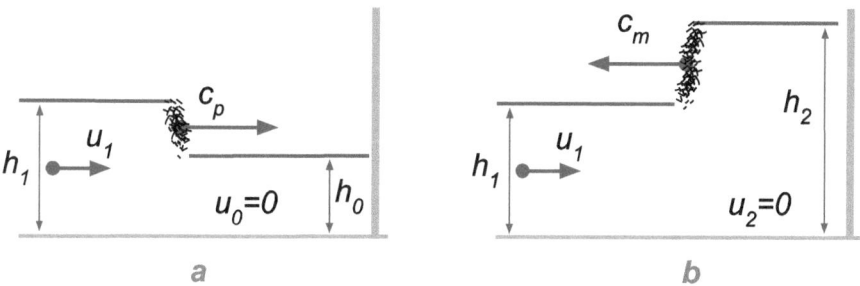

Fig. 7.8 Shock reflection from a wall: (**a**)—an approaching shock, (**b**)—a departing shock

each bore, when the bore was approaching the weir and after it departed, as well as propagation celerities of the incident and reflected waves.

This possibility follows from a simple count of unknowns. An incident shock is characterized by five parameters, which for a channel with a rectangular cross-section are: flow depth and speed in front of the shock h_0 and u_0; those behind the shock h_1 and u_1; and the direct shock celerity c_p (Fig. 7.8a). Parameters of the reflected shock, in the same order, are: h_1 and u_1; h_2 and u_2; and c_m (Fig. 7.8b). Hence the direct and reflected shocks depend on eight independent parameters. Water level measurements provide values h_0 and h_2. Neglecting escape of fluid over the weir, flow velocities by the weir u_0 and u_2 can be assumed zero. The remaining four parameters are constrained by four (two for each shock) conditions.

Stoker (1957) developed a method for computing the height of a reflected shock, given the parameters of the incident one, for a channel with a rectilinear cross-section. Tolkova and Tanaka (2016) adapted Stoker's approach for estimating parameters of an incident shock, given water levels at a barrage, in a channel with an arbitrary shape, as described below.

Following Stoker (1957), let us introduce an auxiliary variable $\zeta = c/u_1$, where c is a celerity of either direct or reflected shock. Then for an incident shock approaching the weir,

$$v_0 = -c = -u_1\zeta, \quad v_1 = u_1 - c = -u_1(\zeta - 1),$$
$$A_0 = A_1 v_1/v_0 = A_1(\zeta - 1)/\zeta, \tag{7.29}$$

with $c = c_p > 0$ being the incident bore celerity. Identical (up to replacing index '0' by index '2') expressions can be written for a reflected shock departing from the weir, with $c = c_m < 0$ being the reflected bore celerity. With due manipulations, substituting (7.29) into (7.26) yields a cubic equation for ζ, function of a dimensionless parameter α and a relative breadth variation before and after the shock:

$$\zeta^3 - \zeta^2 - \alpha\zeta + \frac{\alpha}{2} \cdot \left(1 + \frac{\Delta b}{6b}\right) = 0, \quad \alpha = \frac{gA_1}{bu_1^2} \tag{7.30}$$

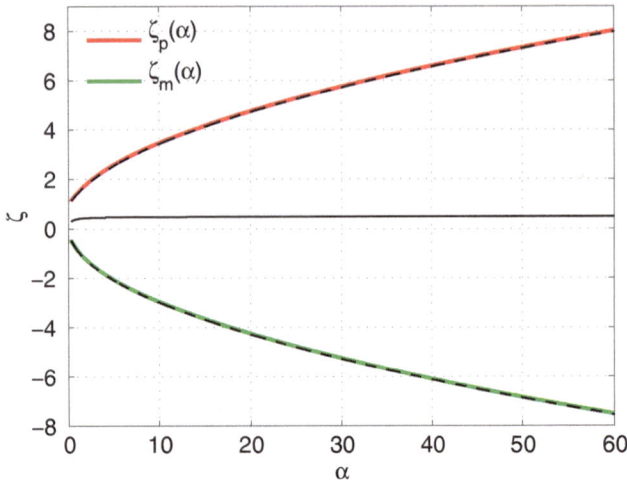

Fig. 7.9 The roots of the cubic equation (7.30) as functions of parameter α, calculated with $\Delta b = 0$ (solid lines) and $\Delta b = 2b/3$ (dashed lines). The largest root (red) determines the relative celerity of a direct shock; the smallest, negative root (green) determines that of a reflected shock; while the remaining root (black) is voided

Note, that $b = (b_1 + b_0)/2$ is an average surface breadth before and after the shock, even though other quantities defining parameter α refer to one side of the shock only. Momentum correction coefficient was set to unity: $m = 1$.

Three real roots of (7.30) as functions of α for $\Delta b = 0$ are plotted in Fig. 7.9 with solid lines. The largest root ζ_p (plotted with a red line) is always greater than unity. This root defines the value c_p/u_1 for the direct shock. The smallest root ζ_m (green line) is always negative, and defines c_m/u_1 for the reflected shock. The remaining root (black line) is positive and does not exceed $1/2$. As seen from (7.29), this last root has no physical meaning, as it would yield a negative flow area.

The tsunami, however, caused significant cross-river inundation and consequent variations of the river breadth (Fig. 7.6). Dashed black lines in Fig. 7.9 show the roots $\zeta_p(\alpha)$ and $\zeta_m(\alpha)$ calculated with $\Delta b = 2b/3$—a breadth variation which corresponds to doubling the river's width across the shock. Nevertheless, the values of $\zeta_p(\alpha)$ and $\zeta_m(\alpha)$ have remained practically unchanged. Therefore, the roots' dependance on Δb can be neglected.

To determine parameter α, we use the last equation in (7.29), which relates functions $\zeta_p(\alpha)$, $\zeta_m(\alpha)$ and the flow areas:

$$A_1/A_0 = \zeta_p/(\zeta_p - 1) = f_1(\alpha) \tag{7.31}$$

$$A_2/A_1 = (\zeta_m - 1)/\zeta_m = f_2(\alpha) \tag{7.32}$$

$$A_2/A_0 = f_1(\alpha) \cdot f_2(\alpha) = f_3(\alpha) \tag{7.33}$$

With A_2 and A_0 obtained from the measurements, the last equation can be solved for α. Then all other shock parameters can be found in a sequence:

$$A_1 = A_0 \cdot f_1(\alpha), \quad u_1 = \sqrt{gA_1/(b\alpha)},$$
$$c_p = u_1 \cdot \zeta_p(\alpha), \quad c_m = u_1 \cdot \zeta_m(\alpha). \tag{7.34}$$

Given the bed profile $d(y)$ at the measurement station (as in Fig. 7.6), the surface elevation η_1 behind an incident shock can be recovered from the flow area A_1.

7.5 How Good Is This Theory?

With this modus operandi, Tolkova and Tanaka (2016) reconstructed the flow conditions near the weir upon arrivals of eleven consecutive tsunami bores. First, they calculated a visible height of each shock above the water surface in front of it: the height of a direct shock $\eta_p = \eta_1 - \eta_0$, and that of a reflected shock $\eta_m = \eta_2 - \eta_1$. Surface levels η_0 and η_2 are directly provided by the water level measurements, whereas η_1 is subject to computations as above. The subsequent heights of eleven shocks at Iino (measured), by the weir (calculated), after reflection from the weir (calculated), and back at Iino (measured) are shown in Fig. 7.10. The reflected shock is always higher than the incident one (Stoker 1957). The shocks reduce in height as they propagate. In this sense, the wave height estimates by the weir are consistent with the observations at Iino for all waves. The reflected waves seem to experience greater dissipation than the direct ones, as suggested both by the present wave height reconstruction, and by the fact that little of the reflected waves were observable at Fukuchi.

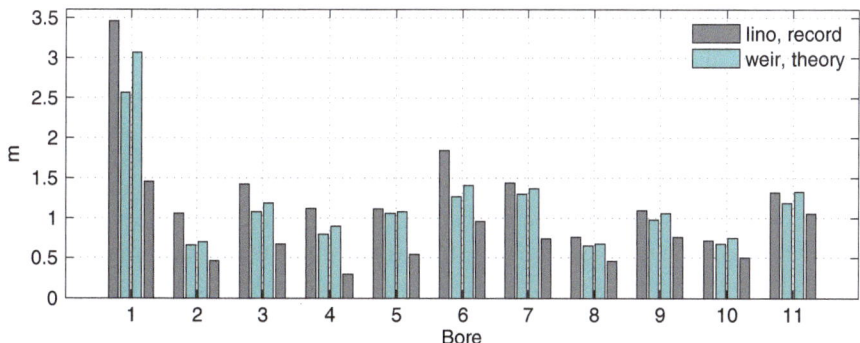

Fig. 7.10 Subsequent heights of the first eleven tsunami bores, measured at Iino and estimated by the weir, as each bore was passing by Iino, approaching the weir, departing from the weir upon the reflection, and returning to Iino

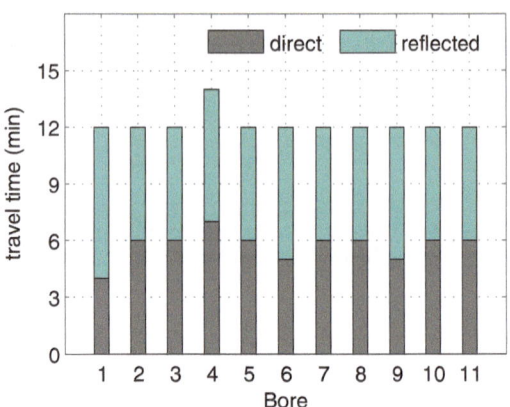

Fig. 7.11 Bores' propagation time (min) from Iino to the weir (direct), and back from the weir to Iino (reflected), as determined from the water level records

A water level record by the weir seemingly does not contain any immediate information about the bore's travel speed. However, the above theory allows to evaluate propagation celerities c_p and c_m of the direct and reflected bores. The calculated celerities can be compared with average propagation speeds from Iino to weir (for the direct waves) or from weir to Iino (for the reflected waves), computed using the records at both location. Each bore's travel time was determined by the arrival times of shock fronts at the locations, as marked by the lower point on a shock in the records at Iino and at the weir (Fig. 1.15). The arrival times are determined with a potential 1-min error, given the 1-min sampling rate in the records.

Travel times to span the Iino-weir distance (2260 m) are shown in Fig. 7.11. On average, the direct bores travel faster than the reflected bores. Interestingly, a bore's travel time for the round trip Iino-weir-Iino remains the same 12 min (with a single outlier, whose presence might be well explained by a time measurement error). The latter circumstance points to the background current as a factor affecting a bore's celerity. The effect of the superposed current (for instance, that set by a previous bore) would be largely nulled on the back-and-forth passage, as observed. Additionally, the presence of a background discharge across the weir would cause a discrepancy between the theoretical and measured celerities, since the theory had employed an assumption of zero current by the weir ($u_0 = u_2 = 0$).

Figure 7.12-left shows the celerities c_p and c_m computed via Eq. (7.34), and average propagation speeds between Iino and the weir for the direct and reflected waves. A fair agreement between the theoretical and measured celerities is observed. A better agreement is observed in Fig. 7.12, right, where theoretical average speeds for the back-and-forth passage $2c_p|c_m|/(c_p + |c_m|)$ are compared with the measured average round-trip celerity of 6.3 m/s. As mentioned, the effect of discharge across the weir, not accounted for in the theory, would be largely reduced with respect to the average round-trip celerities.

As suggested by this exercise, a roaring bore inundating a river's flood plain might conform to equations as simple as those describing a uniform flow in a rectangular channel!

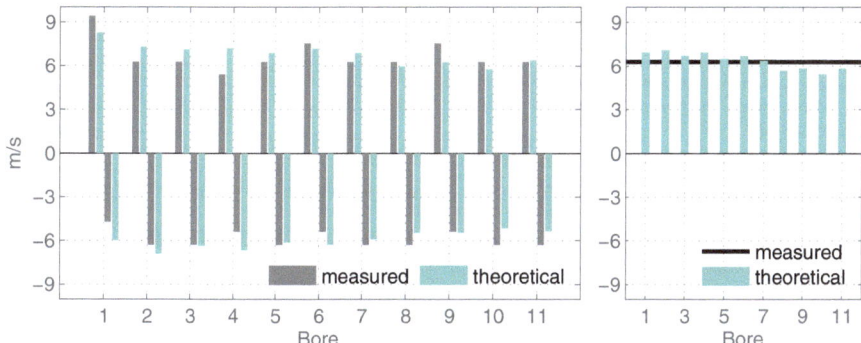

Fig. 7.12 Left: Propagation celerities of the direct and reflected bores obtained with travel time measurements from the Iino and weir records, and theoretically deduced from the water levels at the weir, from Tolkova and Tanaka (2016). Right: Average round-trip celerities obtained with the theoretical estimates c_p and c_m, and that obtained with the travel time measurements

7.6   The Book's Main Points

- Wave penetration into rivers is largely regulated by the wave-induced water accumulation, referred to as the wave set-up or the backwater effect.
- A river's response to an intruding wave is quantified with a new set of parameters characterizing the geometry of the accumulated water.
- The new parameters are studied for their dependance upon the river's bed slope and roughness and the wave's period and amplitude.
- The latter study lays the basis for evaluating a river's response to a tsunami using routine tidal observations.
- A new form of the Shallow-Water Equations and a concept of Wave-Locked Slope (WLS) facilitate interpretation of both tsunami's and tidal propagation in rivers.
- Equations describing a unidirectional wave in a channel with negligible reflection replicate equations describing a steady flow, if the flow is measured by an observer moving with the wave.
- There exists a typical WLS pattern over a tidal cycle, where the surface elevation at low tide tends to take a much steeper upriver path, than that under high tide—a pattern which reveals itself through modulation of a low-amplitude tsunami or seiche propagating in a river atop tide.
- The classical shock-wave theory—intended for representing a uniform flow in a rectangular channel—can be adapted for representing tsunami bores in real-world rivers.

References

Chanson, H. (2010). Undular tidal bores: Basic theory and free-surface characteristics. *Journal of Hydraulic Engineering*, *136*(11), 940–944. https://doi.org/10.1061/(ASCE)HY.1943-7900. 0000264.

Chanson, H. (2012). Momentum considerations in hydraulic jumps and bores. *Journal of Irrigation and Drainage Engineering*, *138*(4), 382–385. https://doi.org/10.1061/(ASCE)IR.1943-4774. 0000409.

Henderson, F. M. (1966). *Open channel flow*. MacMillan Series in Civil Engineering (522 pp.). London: Pearson.

Stoker, J. J. (1957). *Water waves*. New York, NY: Interscience Publishers Inc.

Tolkova, E., & Tanaka, H. (2016). Tsunami bores in Kitakami river. *Pure and Applied Geophysics*, *173*(12), 4039–4054. https://doi.org/10.1007/s00024-016-1351-7.

Tsuji, Y., Yanuma, T., Murata, I., & Fujiwara, C. (1991). Tsunami ascending in rivers as an undular bore. *Natural Hazards*, *4*, 257–266.

Yasuda, H. (2010). One-dimensional study on propagation of tsunami wave in river channels. *Journal of Hydraulic Engineering*, *136*(2), 93–105.

Index

A

Admittance
    factor/function, 94, 98–99
    variations, 99–101
    WLS, 92, 94–98
1964 Alaska tsunami, 10
Arrival time, 10, 118

B

Backwater
    accumulation distance, 52, 60, 62–63
    curve, 29–30, 61–62
    effect, 37
    height, 53, 63
    patterns, 58–61
Bed elevation, 26, 29, 30
    1-D river model, 76
    SWE, 103
    Yoshida River, 42
Bed roughness, 52, 58, 59, 100
Bed slope, 5, 27, 31, 42
Bore
    celerity, 115, 118
    and ordinary wave, 110–112
    propagation time, 118
    shock equations, 103–109
    shock reflection, 113–117
    tsunami waveform, 15–20
    turbulence/ripples, 113
Boundary conditions, 27, 54

C

Cascadia subduction zone, 2
Center for Operational Oceanographic
    Products and Services (COOPS), 4

Channel convergence, 69, 76–78
2010 Chile tsunami
    modulation by tide, 12–14
    traces left by, 2–3
2015 Chile tsunami, 82, 86–88
Cliffs, 53
Columbia River, 4, 7–12, 45–48, 85
Continuity equation, 24, 31, 74
Cross-river integrated shallow-water equations,
    23–28, 71, 103

D

Discharge rate, 29, 74, 76
Dissipation, 11, 33, 52, 64–69, 75–77, 81, 82,
    89, 94, 113, 117
Drag, 24, 26, 32, 72, 79

E

Estuarine tides
    analytical treatment, 78–80
    SWE, 75
    tidal analysis, 48

F

Field survey/observations, 68, 69, 94, 99,
    113
    post-tsunami, 53
    Tanaka's group, 4–5
    wave decay, 63
Flow area, 24, 104, 105, 107, 116, 117
Flow depth, 3, 5, 25, 27, 29, 31, 39, 67, 71,
    78–82, 96, 104, 105, 109–113, 115

© The Author(s) 2018   121
E. Tolkova, *Tsunami Propagation in Tidal Rivers*, SpringerBriefs
in Earth Sciences, https://doi.org/10.1007/978-3-319-73287-9

Flow momentum, 113
Flow velocity, 3, 5, 23, 24, 27, 29, 31, 54, 56,
    58, 66, 76, 82, 100, 104, 109
Frictional convergent channel, 37–39, 75
Friction-convergence balance, 76–78

G
Geometrical focusing, 12

H
Harmonic constituent, 48, 49
High Water path, 79
High water marks, 53, 63, 67–69
Hilbert transform, 89, 90
Hydraulic gradient/slope, 29, 39, 75, 90, 100

I
Ideal estuary
    channel convergence, 77
    1-D river model, 76
    Savenije's condition, 78
    tidal range, 79
2004 Indian Ocean tsunami, 2, 4, 69
Instant wave amplitude, 89, 92

K
Kalu River, 5
Kitakami River, 15, 17, 106, 113

L
Long wave propagation, 23, 71
Low water, 11, 25, 79, 87, 92

M
Mabechi River, 86, 99
Mad River, 15, 16
Manning roughness coefficients, 24, 26, 53, 78
Marigram, 4, 9, 10
Maule River, 3, 12
Method of characteristics, 72
Ministry of Land, Infrastructure,
    Transportation, and Tourism
    (MLIT), 4, 12, 42, 86, 90
Modulation by tide
    2010 Chile tsunami, 12–14

Columbia River, 7–12
    tsunami propagation, 85
Momentum correction coefficient, 25, 105–106
Momentum equation, 24, 31, 76, 79, 81, 104
MSf, 49

N
Naka River, 86, 99
Naruse River, 5, 13, 86, 91
National Oceanic and Atmospheric
    Administration (NOAA), 4, 45
Neap tides, 40
Numerical experiment, 54–55

O
Old Kitakami River, 12–13, 86, 99

P
Phase lag, 27, 28, 78
1964 Prince William's tsunami, 10, 85

R
Riemann invariants, 72, 111
"Ring of Fire," 1
River model, 52
    numerical solutions, 54
    parameters, 54
River stage, 4, 6, 24, 29, 31, 38, 41, 47, 55, 63,
    70, 82

S
Saint-Venant equations, 23
Seiche, 87, 92, 99
Self-registering tide gages, 3, 4
Shallow-water theory, 23, 71
Shock conditions, 103–105
    channel with sloping shores, 107–108
    channel with vertical banks, 106–107
    reflection, 113–117
Spring tides, 40, 49
Stokes flux, 31, 40
Submarine earthquake, 1, 4
Sub-tidal mean river stage (ST-MRS), 40–50
SWE, *see* Shallow-water theory
Synodic cycle, 40, 49

T

Tidal phase, 7, 13, 79, 82, 89, 92, 100, 101
Tidal range, 38, 40, 44, 45, 47, 54, 79, 80
Tidal set-up, 40
    Columbia River, 45–48
    Yoshida River, 42–45
Tide gages
    Beaver and Vancouver records, 10
    Neah Bay and Astoria, 9
    self-registering, 3–4
2011 Tohoku tsunami, 1, 2, 4, 5
    bores, 15–17
    records, 11, 18–19, 80
    time histories, 6
    wave set-up, 44
Tokyo Peil, 4
Traces
    ancient tsunamis, 2
    2010 Chile tsunami, 3
    2015 Chilean tsunami, 86, 88
    2004 Indian Ocean tsunami, 4
Travel time, 9, 10, 92, 96, 118, 119
Turbulence, 113
Turbulent bore, 15

U

Undular bore, 15, 113

W

Water accumulation, 4–6
Water level measurements, 4, 86–87
Wave amplitude, 31, 33, 53
    backwater, 58, 59, 63
    high water marks, 67, 68
    WLSs, 92, 96, 99
Wave breaking, 111
Wave dissipation, 33, 64, 67, 76, 82
Wave-Locked Slope (WLS), 73, 89–90
    admittance computations, 94–98
    admittance variations, 99–101
    instant Tsunami amplitude, 92–94
Wave period, 27, 28, 31, 32, 40, 52, 59, 63, 67–69
Wave set-up, 30, 34
    tidal observations, 40–50
WLS, *see* Wave-Locked Slope (WLS)

Y

Yoshida River, 5, 12, 42–45, 86